Natural Capital and Climate Smart Agriculture

India is the fastest growing and the world's third-largest economy in terms of GDP in PPP terms. Sustainable development of India will ensure the welfare of the inhabitants of this most populated country. This book assesses trends of natural capital and areas of improvement through climate-resilient agricultural adaptation in India.

The book looks at how the agricultural sector can become more climate resilient to ensure food security and human capital development. It also suggests a policy framework towards climate-resilient agricultural development. It outlines determinants of climate-smart agricultural practices and their impact on agricultural yield, biodiversity, and food security, as well as outreach activities for wider collaboration from around the world.

This book will interest those who are researching accounting natural capital impacts of climate-resilient agriculture and 2030 SDGs.

Pradyot Ranjan Jena is a Professor at the National Institute of Technology Karnataka, Surathkal, India.

Shunsuke Managi is a Distinguished Professor of Technology and Policy and Director of the Urban Institute at Kyushu University, Japan.

Ritanjali Majhi is a Professor at the National Institute of Technology Karnataka, Surathkal, India.

Routledge Studies in Development Economics

For more information about this series, please visit: www.routledge.com/
Routledge-Studies-in-Development-Economics/book-series/SE0266

Natural Capital and Climate Smart Agriculture

Measuring Progress towards Sustainability and Policy Making in India

Pradyot Ranjan Jena, Shunsuke Managi, and Ritanjali Majhi

LONDON AND NEW YORK

First published 2025
by Routledge
4 Park Square, Milton Park, Abingdon, Oxon OX14 4RN

and by Routledge
605 Third Avenue, New York, NY 10158

Routledge is an imprint of the Taylor & Francis Group, an informa business

British Library Cataloguing-in-Publication Data
A catalogue record for this book is available from the British Library

ISBN: 9781032269047 (hbk)
ISBN: 9781032269054 (pbk)
ISBN: 9781003290445 (ebk)

DOI: 10.4324/9781003290445

Typeset in Galliard
by KnowledgeWorks Global Ltd.

Contents

About the authors

Pradyot Ranjan Jena is a Professor at the National Institute of Technology Karnataka, Surathkal. Prof. Pradyot Ranjan Jena has a distinguished career in Economics with over two decades of experience in the fields of Development Economics, Agricultural Economics, and Environmental Economics. He holds a Ph.D. from the prestigious Indian Institute of Technology (IIT) Kanpur and has built a career that bridges academic research, policy analysis, and on-the-ground development initiatives. His extensive research, funded by leading international organizations such as the Asian Development Bank Institute (ADBI) and CGIAR, has explored critical issues ranging from climate-smart agriculture to the socio-economic impacts of certification schemes in the agri-food sector. His work has been instrumental in driving sustainable development policies in India and beyond. Prof. Jena's scholarly contributions include over 60 peer-reviewed journal articles, numerous book chapters, and several edited volumes. His research has been widely published in high-impact journals, where he has explored topics such as the integration of financial markets in India, the environmental implications of trade liberalization, and the adoption of climate-resilient agricultural practices. In addition to his academic achievements, Dr. Jena has played a significant role in international development, particularly through his work with the International Maize and Wheat Improvement Center (CIMMYT) in Nairobi. There, he led initiatives aimed at eradicating poverty and hunger in Sub-Saharan Africa by innovating advanced agricultural technologies.

Shunsuke Managi is a Distinguished Professor of Technology and Policy & Director of the Urban Institute at Kyushu University, Japan. He is a director for UN Inclusive Wealth Report 2018 (IWR 2018) and IWR2023, ISC working group member for UN' The Global Sustainable Development Report 2023, a lead author for the IPCC, a coordinating lead author for the IPBES, a coordinating lead author UNESCO Education Assessment, lead author of Policy Briefs T20 for the Presidency of G20 in 2022, 2023, an editor of "Economics of Disasters and Climate Change", "Environmental Economics and Policy Studies", "Encyclopedia of Energy, Natural Resource, and Environmental Economics, 2e", "The Routledge Handbook of Environmental Economics in Asia". He is the co-chair of the Scientific Committee of the 2018 World Congress of Environmental and Resource Economists. In 2020 he was elected as a council member of the Science Council of Japan.

Ritanjali Majhi is a Professor at the National Institute of Technology Karnataka (NITK). Dr Majhi has made significant contributions to the field of Analytical Marketing, holding a PhD from the prestigious Birla Institute of Technology (2009). Her research expertise encompasses a wide array of subjects including Consumer Decision Making, Time Series Prediction, Artificial Intelligence, and the application of soft and evolutionary techniques to Management. Dr Majhi's scholarly work is further enriched by her collaborations with esteemed institutions like Liverpool University and Sheffield University in the UK. Dr Majhi has spearheaded several impactful research projects, notably in areas such as circular agri-food systems, soil health and productivity, household waste management, and climate-smart agriculture practices. Her projects have attracted substantial funding from organizations like SPARC, the Government of India, the Asian Development Bank Institute, and ICSSR, reflecting the critical importance and innovative nature of her work. In addition to her research, Dr. Majhi actively contributes to academic governance, serving on boards of studies at various institutions including the Department of Management at NIT Andhra Pradesh and the MSC Business Analytics programme at Manipal University. Dr. Majhi's prolific writing includes numerous publications in leading international journals such as Elsevier, Springer, and the Journal of Public Affairs. A recognized authority in her field, Dr. Majhi has contributed to several book chapters and is the editor of the book "Advances in Intelligent Systems and Computing", published by Springer.

Preface

The concept of "sustainability" is no longer confined to the high-decibel IPCC conventions or the scientific laboratories; it has become a matter of conscious choice in every household. As the global mean temperature has breached the proverbial 1.5° C above the pre-industrial revolution temperature on some days of the year 2023, intense heat waves are experienced in many parts of the world. Melting of Antarctic glaciers, rising sea levels, frequent flooding of cities, and rapidly vanishing rainforests have sounded alarm bells for the survival of humankind, flora, and fauna. Sustaining biodiversity is crucial for maintaining the ecosystem balance. Natural capital which includes forests, waterbodies, landmass, flora, and fauna is the key element of biodiversity and ecosystem balance. Several scientifically acclaimed studies have reported that natural capital in many developing countries is in a state of decline. This does not bode well for achieving sustainable development. Dwindling natural capital severely destabilizes the ecological balance and creates survival threats for living organisms. Agricultural production is an important function of natural capital and the changes in the agricultural production system influence the state of natural capital significantly. Against this backdrop, this book is an attempt to understand the linkage between sustainable agricultural practices and natural capital preservation in the Indian context.

The global population is projected to rise to 9 billion by 2050. Producing adequate and nutritious food for such a large population poses a significant challenge to the agri-food system. Technological development in the agriculture sector is imperative to achieve this daunting task. The recent episodes of extreme climate shocks have magnified this challenge by reducing the adaptive capability of the agriculture sector worldwide. Severe heatwaves spurred by El Nino events and erratic rainfall have created enormous losses for farmers in both the developed and developing countries, but more so for the latter as the farmers in these countries have been severely affected due to their lack of adaptive capacity. In this context, climate-smart agriculture (CSA) emerges as a potential solution to enhance the adaptive capacity of small and marginal farmers and mitigate the impacts of climate change.

CSA technology is a basket of modern and traditional agricultural technologies that have scientifically proven benefits for soil ecosystem conservation and greenhouse gas (GHG) mitigation. Modern CSA techniques include the application of precision agricultural tools such as laser-guided land levelling, sensor-guided application of fertilizer and pesticides, drip irrigation, minimum tillage, and soil testing. When these techniques are combined with traditional sustainable methods such as crop rotation, mulching, row planting, and crop diversification, beneficial natural capital outcomes are generated. Precision agriculture tools prevent excessive water use, mitigate emissions of GHGs, reduce soil contamination, and enhance soil nutrition. So, CSA has the potential to produce sustainable agricultural growth that not only improves the productivity of crops and thereby higher incomes but also improves natural capital and biodiversity.

Throughout this book, this linkage is established based on empirical evidence from India. The empirical analysis presented in the book is based on both historical climate data and household survey-based socio-economic data from different Indian states. In different chapters, we delved into the nuances of the natural capital-sustainable agriculture linkage. The time-series trend of natural capital components such as agricultural land use, change in forest stock, carbon capture, mineral stock, and renewable energy generation are presented to show the direction of change in natural capital stock. In Chapter 2, rigorous econometric techniques are used to measure the impact of climate change on agricultural production without suitable adaptation. A vulnerability index is constructed by combining the climatic and socio-economic parameters using district-level time-series data to show the urgency of climate change adaptation. In subsequent chapters, the concept and function of CSA are explained, and how it has the potential to achieve the triple objectives of higher farm productivity, adaptation to climate change, and conservation of natural capital has been discussed. Chapter 5 is dedicated to exploring an important social aspect of this sustainability debate, which is gender relations. The role of women in agriculture has been well established from past research; however, the role of women in adopting and practising CSA has a fascinating side that has been observed from our several field visits and focus group discussions. Although the CSA concept has been presented as a potential solution for climate change woes in agriculture and biodiversity, an increasingly emerging issue of concern has been that the adoption rate among the small and marginal farmers that occupy around 70% of the farming population in developing countries has been slow. In countries such as India, Brazil, Vietnam, and Thailand monocropping is prevalent. CSA practices such as crop diversification or crop rotation are less preferred due to an apparent lack of economic profits. Therefore, rigorous empirical investigation is needed to ascertain the economic benefits of CSA for their wider adoption. Chapter 6 undertakes a systematic impact evaluation of some of the CSA practices in the eastern part of India. Finally, Chapters 7 and 8 show the intricate relationship between natural capital and sustainable agriculture. There are both short-term and long-term interactions between certain farming practices and ecosystem functions. Understanding this complex relationship helps design effective policies.

This book has useful material for students of environmental economics, PhD scholars, researchers, practitioners, and policymakers. The case studies used throughout the book have produced many policy-relevant insights that can be discussed, debated, and researched further. Both the empirical methodology used and the results are expected to draw significant attention from academia, practitioners, and policymakers. The secondary and primary datasets used in preparing the chapters are current, some of them are collected during the COVID-19 episodes in India and after it. By the time this book gets published, India and much of the world would have experienced an unprecedented level of heatwave causing fatality among humans, animals, and birds. There has been a heightened awareness and shock as to how the globe is heating up, which requires a drastic behavioural change in people's actions. Common citizens, the business world, and politicians must take this threat as real and get their acts together to achieve sustainability. This book aims to act as a catalyst in that endeavour.

Pradyot Ranjan Jena
Professor National Institute of Technology Karnataka,
Surathkal, India

Acknowledgements

Our sincere gratitude goes to the institutions where we currently work, the National Institute of Technology Karnataka, Surathkal and Kyushu University, for supporting our work in preparing this book manuscript. Several people deserve our appreciation for the completion of this volume. Foremost among them are the respondents of our household surveys, the farming households of Odisha who waited patiently and kindly provided us with the data that are used in various chapters of this book. We are thankful to the data enumerators who have toiled hard in the field even braving the risk of COVID-19 infection to collect the data. Appreciations are due to the agricultural extension officers, village-level workers (VLWs), agricultural officers at the state agricultural department, and the agricultural scientists who have generously supported our work. We are thankful to the Ministry of Education (MoE) of the government of India for partially supporting us with a research grant.

We sincerely thank Dr Purna Chandra Tanti, Dr Sunil Khosla, Mr Kamalakanta Datta and Mr Sunil Meher, our previous and current PhD research scholars at NITK Surathkal, for their excellent editing support. Finally, we thank the participants of various conferences where we presented our research work for their valuable insights to improve the overall quality of the content used in this book.

1 Natural Capital in India

An Overview of Trends and Status of Natural Capital Stock

1.1 Natural Capital: A Brief History

Originally coined by early economists to explain how humans and the environment are inextricably linked, natural capital has its roots in early economics and the concept of environmental sustainability. The idea of natural capital was first introduced by economist E.F. Schumacher in the 1970s, who recognized the importance of considering the value of natural resources in economic decision-making. As concerns about natural resource depletion and the influence of human activities on the environment grew in the 1990s, the concept gained broader acceptance.

Since then, there have been numerous international agreements and initiatives that incorporate natural capital, including the United Nations Conference on Environment and Development (UNCED) and the Millennium Ecosystem Assessment. The concept has also been integrated into mainstream economic thinking and is increasingly being used as a framework for evaluating the trade-offs between economic development and environmental preservation.

Today, the concept of natural capital continues to evolve as new research and understanding of the relationships between the natural world and human well-being emerge. The concept is seen as a key component of sustainable development and is increasingly being used by businesses, governments, and individuals to make decisions about the use and management of natural resources. Natural capital is often seen as a critical component of economic growth and development. Natural resources provide the raw materials for many industries and are the basis for many economic activities, such as agriculture, forestry, and fishing. The exploitation of natural capital can lead to short-term economic benefits, such as increased employment and income.

However, if natural capital is depleted or degraded, it can have negative impacts on the economy in the long term. For example, if forests are destroyed or water resources are polluted, the resulting damage to ecosystem services can reduce the productivity of agriculture, fishing, and other industries that depend on these resources. The degradation of natural capital can also lead to increased costs for environmental remediation and the provision of alternative resources, reducing the overall benefits of economic development.

DOI: 10.4324/9781003290445-1

Therefore, the relationship between natural capital and economic growth is complex and depends on the manner in which natural capital is managed and utilized. In order to achieve sustainable economic growth, it is necessary to balance the exploitation of natural capital with its preservation and restoration. This requires a more integrated and holistic approach to decision-making, considering the interdependence between the economy and the natural environment.

Natural capital plays an important role in regulating carbon emissions and mitigating the impacts of climate change. Forests, wetlands, and other ecosystems act as carbon sinks, absorbing and storing carbon dioxide from the atmosphere. In addition, natural systems provide important ecosystem services, such as water filtration and air purification, that help to mitigate the impacts of climate change and air pollution.

However, the exploitation and degradation of natural capital can also contribute to increased carbon emissions. For example, deforestation, the destruction of wetlands, and other forms of land use change can release large amounts of carbon dioxide into the atmosphere. In addition, the extraction and burning of fossil fuels, such as coal, oil, and natural gas, are major sources of greenhouse gas emissions, contributing to global warming and climate change.

Therefore, the relationship between natural capital and carbon emissions is complex and requires a balanced approach to resource management. To mitigate the impacts of climate change, it is important to reduce carbon emissions from human activities and to protect and restore natural systems that provide important ecosystem services. This requires a more integrated and holistic approach to decision-making, considering the interdependence between the economy and the natural environment.

1.2   Natural Capital: Indian Context

India has a rich and diverse natural capital stock, including forests, wetlands, and a range of ecosystems that support a variety of plant and animal species. Natural capital in India is important for a number of reasons, including its contribution to the economy, its role in regulating the climate and water cycles, and its support for human health and well-being.

However, India's natural capital is facing significant challenges, including deforestation, land use change, and pollution. These pressures are driven by a growing population, rapid economic development, and increasing demands for natural resources. As a result, many of India's ecosystems and natural resources are being degraded, reducing their ability to provide valuable services and support human well-being.

In recent years, the Indian government has taken steps to address these challenges and protect its natural capital. This includes the implementation of policies and programmes aimed at improving natural resource management, conserving biodiversity, and reducing the impact of human activities on the environment. In addition, the government has increased investment in renewable energy, recognizing the importance of transitioning to a more sustainable and low-carbon development path.

Overall, India's natural capital is of critical importance for its economy, environment, and human well-being. Protecting and restoring this resource will be critical for ensuring sustainable development and a secure future for the country and its people. This chapter presents a holistic picture of the current position of different natural capital components in India. Natural capital can be classified into two primary categories: (1) renewable resources and (2) non-renewable resources. Renewable resources are further subdivided into (a) forest resources, encompassing tree cover; (b) fisheries, represented by fish production; (c) agricultural land, which includes cropland and pastureland; and (d) renewable energy resources, denoted by energy produced from renewable sources. In contrast, non-renewable resources can be categorized into (e) coal resources.

1.3 Agricultural Land Holdings

Land holdings in India assume a pivotal role in the nation's agricultural sector and rural economy. With a substantial proportion of the population engaged in agriculture, land serves as a valuable asset and a fundamental source of livelihood for millions of farmers. The land holdings across India exhibit a diverse array of sizes, ownership patterns, and land use practices. While small and marginal holdings predominantly characterize the landscape, there also exist larger land holdings owned by affluent farmers and agricultural corporations. The fragmented nature of land parcels, arising from inheritance laws and population growth, poses challenges to agricultural productivity and endeavours towards land consolidation. Moreover, tenancy arrangements significantly contribute to agricultural production, with tenant farmers assuming a vital role in the sector. The Indian government has proactively implemented a range of initiatives and land reforms aimed at addressing concerns of equitable land distribution, supporting small and marginal farmers, and fostering agricultural development. Gaining a comprehensive understanding of India's land holdings is imperative for comprehending the dynamics of the agricultural sector and the socioeconomic fabric of the country. The relentless growth of India's population by over 2% has brought to the forefront a pressing demand for food to alleviate hunger. This, in turn, indirectly stimulated the expansion of the Net sown area or Arable area. In the decade from 1950 to 1960, there was an increase in net area sown amounting to 14,193 thousand hectares (ha), while the same for total area sown was 20,931 thousand ha, indicating an increase of 6,738 thousand ha of area sown more than once during that period resulting in a cropping intensity of 3.9% (see Table 1.1). During this period, there was a decrease in agricultural land with a magnitude of 8,420 thousand ha, and an increase in cultivated land area of 15,439 thousand ha. The subsequent decade, from 1960 to 1970, saw a dramatic reduction in the net area sown to 5,496 thousand ha, with the change in total area sown recorded

Table 1.1 Decadal Changes in Land Use in India (1950–1951 to 2021–2022) in '000 ha

Year	Change in net area sown	Change in gross area cropped	Change in area sown more than once	Change in agricultural land	Change in cultivated land	Change in cropping intensity (in %)
1950–1960	14,193	20,931	6,738	–8,420	15,439	3.9
1960–1970	5,496	9,493	3,997	1,159	6,180	2.3
1970–1980	–1,960	3,798	5,758	2,756	3,140	4.4
1980–1990	2,051	9,639	7,588	55	919	5
1990–2000	–1,807	2,654	4,461	–1,314	–594	3.6
2000–2010]	–2,163	3,848	6,010	–1,276	–931	4.8
2010–2020	–1,469	13,231	14,699	–1,924	–2,078	11
2020–2022	–537	3,051	3,589	–154	–268	2.7

Source: Ministry of Agriculture & Farmers Welfare, Govt. of India (ON3410 & past issues). Data Source: indiastat.com. Data accessed on 25 July 2023.

at 9,493 thousand ha. The differential between total and net area sown was 3,997 thousand ha. There was an increase in agricultural land by 1,159 thousand ha, while cultivated land expanded to 6,180 thousand ha, reflecting a cropping intensity of 2.3%. The 1970–1980 period marked a significant negative shift with a net area sown of –1,960 thousand ha, despite a total area sown of 3,798 thousand ha. During this time, agricultural land increased by 5,758 thousand ha, and cultivated land grew by 2,756 thousand ha, leading to a cropping intensity of 4.4%. From 1980 to 1990, the net area sown improved to 2,051 thousand ha, and the total area sown increased by 9,639 thousand ha. There was a substantial increase in agricultural land by 7,588 thousand ha. However, the cultivated land saw a minimal increase of only 55 thousand ha, which resulted in a cropping intensity of 5%. During the 1990–2000 decade, the net area sown fell to –1,807 thousand ha, with a total area sown of 2,654 thousand ha. Agricultural land increased by 4,461 thousand ha, but cultivated land decreased by –1,314 thousand ha. The cropping intensity for this period was 3.6%. From 2000 to 2010, the net area sown continued to decline to –2,163 thousand ha, while the total area sown increased to 3,848 thousand ha. Agricultural land saw an increase of 6,010 thousand ha, but cultivated land experienced a reduction of –1,276 thousand ha. This period observed a cropping intensity rise to 4.8%. In the decade from 2010 to 2020, the net area sown was –1,469 thousand ha, with a substantial increase in the total area sown to 13,231 thousand ha. Agricultural land grew significantly by 14,699 thousand ha, although cultivated land decreased by –1,924 thousand ha, resulting in a high cropping intensity of 11%. The most recent data from 2020 to 2022 indicates a change in net area sown of –537 thousand ha, with an increase in total area sown of 3,051 thousand ha. Agricultural land increased by 3,589 thousand ha, while cultivated land saw a slight reduction of –154 thousand ha. The cropping intensity for this period was 2.7% (Figure 1.1).

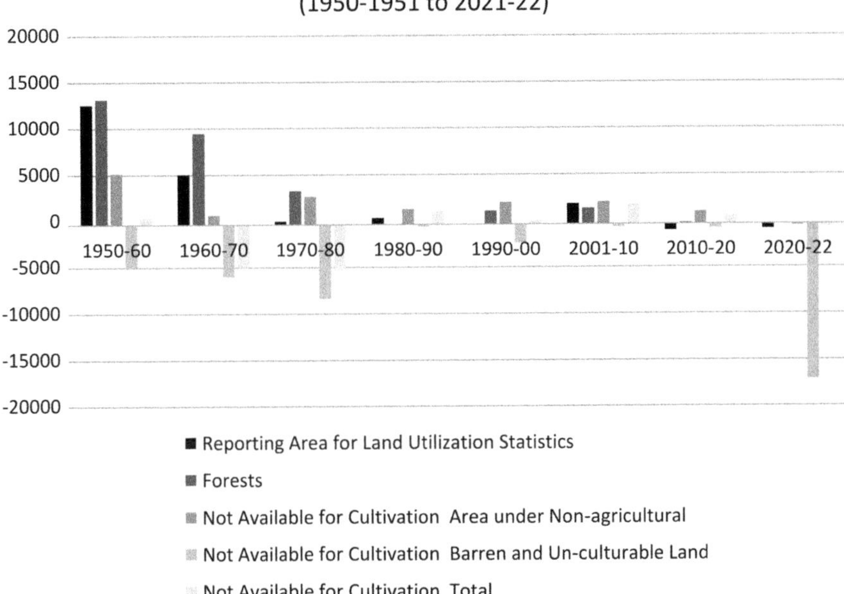

Figure 1.1 Decadal Change in Land Use Classification in India (1950–1951 to 2021–2022).

A notable observation can be derived from Table 1.2 on the varying extent of Net Area Sown across different states and union territories (UTs). This variation highlights the diverse agricultural practices and land utilization strategies adopted across the country. For instance, the state of Rajasthan exhibits the highest Net Area Sown at 18,130 thousand ha, followed closely by Maharashtra with 16,590 thousand ha, Uttar Pradesh with 16,096 thousand ha, and Madhya Pradesh with 15,823. Conversely, UTs like Chandigarh, Lakshadweep, and Ladakh have negligible Net Area Sown, indicating limited agricultural activity. Comparing the Net Area Sown to the Total Cropped Area provides insight into the intensity of cropping practices. Dadra and Nagar Haveli and Daman and Diu, Lakshadweep, and Manipur1 demonstrate the highest efficiency, utilizing 100% of their net sown area for cropping. Other regions like Ladakh at 90.91%, Goa at 88.19%, and Nagaland at 83.07% also show high land use intensity. In contrast, states like Punjab (51.96%), Madhya Pradesh (52.66%), and Tripura (52.36%) exhibit lower percentages, indicating less intensive agricultural practices. The national average percentage of the total cropped area to the net sown area is 64.34%, suggesting that on average, about two-thirds of the net sown area is utilized for cropping in India. The total area sown more than once in India amounts to 78,152 ha. Madhya Pradesh

Table 1.2 State-wise Land Use Classification in India (2021–2022) (in '000 ha)

States/UTs	Net area sown	Total cropped area	Area sown more than once	Agri. land/ cultivable land/ culturable land/arable land	Cultivated land	Un-cultivable land	Un-cultivated land
Andaman & Nicobar Islands	15	39	25	28	17	687	697
Andhra Pradesh	6,038	7,328	1,290	8,987	7,537	7,309	8,759
Arunachal Pradesh	243	338	95	431	276	6,723	6,878
Assam	2,749	3,872	1,123	3,321	2,863	4,510	4,968
Bihar	5,070	7,329	2,258	6,542	6,073	2,818	3,286
Chandigarh	1	2	1	1	1	8	8
Chhattisgarh	4,631	5,704	1,073	5,552	4,911	8,538	9,180
Dadra and Nagar Haveli and Daman and Diu	23	23	0	23	23	24	25
Delhi	22	58	37	53	34	113	132
Goa	127	144	18	141	131	221	230
Gujarat	9,720	14,759	5,038	12,428	10,381	6,382	8,429
Haryana	3,611	6,566	2,955	3,847	3,696	524	676
Himachal Pradesh	528	890	362	831	605	3,746	3,972
Jammu & Kashmir	733	1,134	401	1,075	821	2,768	3,023
Jharkhand	1,379	1,845	467	4,324	2,747	3,646	5,223
Karnataka	11,166	14,748	3,582	12,836	11,844	6,214	7,206
Kerala	2,029	2,523	494	2,223	2,083	1,663	1,803
Ladakh	20	22	1	28	21	264	271
Lakshadweep	2	2	0	2	2	3	3
Madhya Pradesh	15,823	30,049	14,226	17,432	16,167	13,324	14,590
Maharashtra	16,590	25,730	9,140	20,466	18,046	10,292	12,712
Manipur	393	393	–	399	393	1,688	1,695
Meghalaya	269	325	56	1,015	323	1,189	1,881
Mizoram	145	206	62	367	191	1,672	1,848
Nagaland	265	319	54	672	357	984	1,300
Odisha	4,322	4,997	676	6,782	5,265	8,063	9,580
Puducherry	16	28	12	28	20	21	29
Punjab	4,113	7,916	3,803	4,225	4,203	807	830
Rajasthan	18,130	27,442	9,312	25,463	19,715	8,829	14,578
Sikkim	77	140	63	97	84	344	357
Tamil Nadu	4,909	6,348	1,439	8,105	5,709	4,928	7,324
Telangana	5,625	8,026	2,401	6,715	6,178	4,493	5,030
Tripura	255	487	231	270	256	779	793
Uttar Pradesh	16,096	28,200	12,104	18,264	17,062	5,250	6,451
Uttarakhand	594	969	376	1,541	691	4,460	5,310
West Bengal	5,281	10,259	4,978	5,595	5,535	3,089	3,149
India	**141,007**	**219,158**	**78,152**	**180,112**	**154,262**	**126,374**	**152,224**

Source: Ministry of Agriculture & Farmers Welfare, Govt. of India (ON2807).
Data Source: indiastat.com. Data accessed on 25 July 2023.

leads with the highest area sown more than once, covering 14,226 ha, followed by Uttar Pradesh with 12,104 ha and Rajasthan with 9,312 ha. Other states showing significant areas include Maharashtra with (9,140 ha), Gujarat (5,038 ha), and Karnataka (3,582 ha). In contrast, states like Lakshadweep, Dadra and Nagar Haveli, Daman and Diu, and Manipur reported negligible or zero ha under this category. Overall, India has a total of 180,112 ha of Agri. Land/Cultivable Land, out of which 154,262 ha are currently under cultivation. The un-cultivable and un-cultivated lands amount to 126,374 ha and 152,224 ha, respectively.

Further insights into the dynamic nature of agricultural practices can be gleaned from Table 1.3, which shows data on the net sown area in India from 2008–2009 to 2020–2021 across states and UTs. Several states exhibited notable changes: Andhra Pradesh experienced a substantial decrease from 10,868,000 ha to 5,915,000 ha, marking a 45.6% decline. Similarly, Maharashtra saw a slight reduction from 17,422,000 ha to 16,650,000 ha, indicating a 4.4% decrease. In contrast, Karnataka demonstrated growth with an increase from 10,174,000 ha to 11,453,000 ha, reflecting a rise of 12.5%. Telangana showed significant improvement, increasing its net sown area from 4,377,000 ha to 5,927,000 ha, a notable 35.3% increase. Overall, India's net sown area experienced a marginal decline from 141,899,000 ha to 141,544,000 ha, reflecting a 0.2% decrease nationwide.

Figure 1.2 shows the amount of land dedicated to each cereal crop in million ha, the total production output in million tonnes, and the yield per ha in kilos. Rice and wheat are the most important cereal crops in terms of both production and yield, with rice occupying 45.07 million ha and yielding

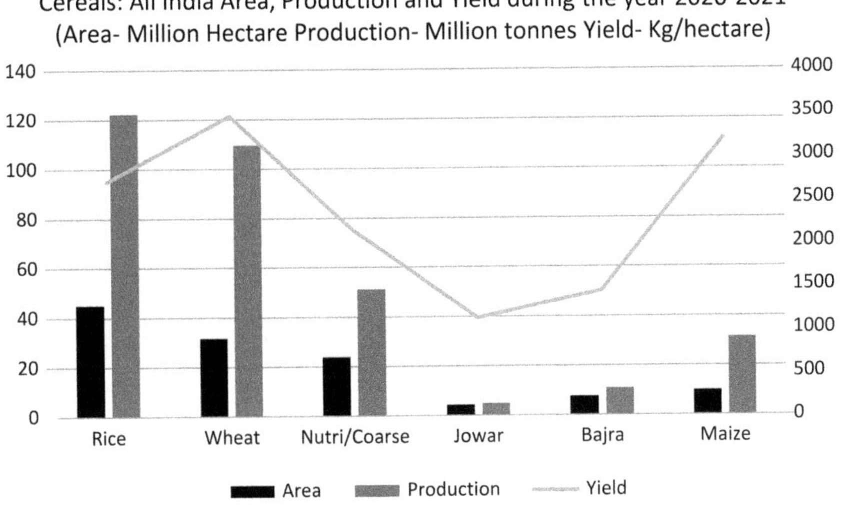

Figure 1.2 Cereals: All India Area, Production and Yield during the Year 2020–2021.

Table 1.3 State-wise Net Sown Area in India (2008–2009 to 2020–2021) (in '000 ha)

States/UTs	2008–2009	2009–2010	2010–2011	2011–2012	2012–2013	2013–2014	2014–2015	2015–2016	2016–2017	2017–2018	2018–2019	2019–2020	2020–2021
Andhra Pradesh	10,868	9,991	11,186	11,161	11,117	6,448	6,236	6,209	6,077	6,048	6,049	5,884	5,915
Arunachal Pradesh	211	212	213	215	216	225	225	227	232	233	234	235	242
Assam	2,810	2,811	2,811	2,811	2,809	2,820	2,827	2,801	2,774	2,723	2,723	2,699	2,724
Bihar	5,554	5,332	5,259	5,396	5,402	5,252	5,278	5,205	5,293	5,242	5,167	5,077	5,045
Chhattisgarh	4,710	4,683	4,697	4,677	4,671	4,686	4,681	4,651	4,664	4,653	4,679	4,635	4,623
Delhi	23	22	22	22	22	22	22	22	22	22	22	22	22
Goa	135	132	131	132	132	129	129	130	130	129	128	127	127
Gujarat	9,974	9,902	10,302	10,114	10,114[1]	10,114[1]	9,618	9,770	9,886	9,891	9,390	9,787	9,822
Haryana	3,576	3,550	3,518	3,513	3,513	3,497	3,522	3,522	3,499	3,517	3,601	3,552	3,611
Himachal Pradesh	539	538	543	550	548	549	548	551	548	543	542	530	526
Jammu & Kashmir	739	735	732	746	745	741	758	754	757	752	713	720	736
Jharkhand	1,504	1,250	1,085	1,250	1,406	1,384	1,385	1,386	1,451	1,444	1,281	1,291	1,328
Karnataka	10,174	10,404	10,523	9,941	9,793	9,923	10,044	10,006	9,855	9,895	10,664	10,804	11,453
Kerala	2,089	2,079	2,072	2,040	2,048	2,051	2,043	2,023	2,015	2,040	2,034	2,026	2,035
Madhya Pradesh	14,941	14,972	15,119	15,237	15,352	15,422	15,351	15,149	15,228	15,191	15,205	15,512	15,800
Maharashtra	17,422	17,401	17,406	17,386	17,344[1]	17,368[1]	17,345[1]	17,192[1]	16,910[1]	16,943[1]	16,815[1]	16,722[1]	16,650[1]
Manipur	236	234	348	365	309[1]	377[1]	383[1]	437[1]	469[1]	446[1]	441[1]	331[1]	410[1]

(Continued)

Table 1.3 (Continued)

States/UTs	2008–2009	2009–2010	2010–2011	2011–2012	2012–2013	2013–2014	2014–2015	2015–2016	2016–2017	2017–2018	2018–2019	2019–2020	2020–2021
Meghalaya	284	283	284	285	285	286	286	245	252	253	255	255	253
Mizoram	95	123	121	97	116	114	145	145[1]	145[1]	145[1]	145[1]	145[1]	145[1]
Nagaland	316	361	362	379	380	380	384	384	385	385	384	384	386
Odisha	5,604	4,761	4,682	4,394	4,386	4,495	4,474	4,198	4,099	3,863	4,006	4,102	4,179
Puducherry	19	19	19	18	16	15	16	15	15	15	15	15	15
Punjab	4,169	4,158	4,158	4,134	4,150	4,145	4,119	4,137	4,130	4,124	4,119	4,127	4,126
Rajasthan	17,551	16,974	18,349	18,034	17,479	18,268	17,521	18,024	18,169	17,903	17,778	18,032	17,948
Sikkim	77	77	77	77	77[1]	77[1]	77[1]	77[1]	77[1]	77[1]	77[1]	77[1]	77[1]
Tamil Nadu	5,043	4,892	4,954	4,986	4,544	4,714	4,819	4,833	4,347	4,639	4,582	4,738	4,833
Telangana	–	–	–	–	–	4,961	4,377	4,175	4,774	4,898	4,660	5,500	5,927
Tripura	256	256	255	255	255	255	255	255	255	255	255	256	255
Uttar Pradesh	16,562	16,589	16,593	16,623	16,564	16,546	16,598	16,469	16,564	16,542	16,538	16,368	16,368 (P)
Uttarakhand	754	737	723	714	706	701	700	698	691	673	648	638	621
West Bengal	5,294	5,256	4,981	5,198	5,205	5,234	5,238	5,243	5,246	5,247	5,248	5,250	5,282

Source: Reserve Bank of India (ON2151). Data Source: indiastat.com. Data accessed on 25 July 2023.

122.27 million tonnes at a yield of 2,713 kg/ha, and wheat occupying 31.61 million ha and yielding 109.52 million tonnes at a yield of 3,464 kg/ha. These two mainstays contribute greatly to India's food output. Furthermore, India's complex agricultural landscape, with different cereals such as jowar, bajra, and maize each having distinct area, production, and yield characteristics, reflecting the country's numerous agroclimatic zones and agricultural techniques. The aggregate data for all grains, which totalled 122.18 million ha, 330.09 million tonnes, and a yield of 14,082 kg/ha, demonstrates cereals' significant significance in India's agricultural output over the defined time.

1.4   Forest and Fisheries

The forests and fisheries of India, which hold immense significance as integral elements of its rich natural heritage, occupy a pivotal position within the nation's ecological framework. The current condition of these invaluable resources is emblematic of a delicate equilibrium between abundance and formidable challenges. India's expansive forest cover, encompassing diverse ecosystems and harbouring a vast array of plant and animal species, serves as a testament to both the marvels of the natural world and the profound impacts of human activities, climate change, and habitat fragmentation. Similarly, the fisheries of the country, spanning expansive marine and freshwater domains, play a vital role in providing sustenance and livelihoods to numerous communities. However, they also confront persistent pressures stemming from issues such as overfishing and environmental degradation. A comprehensive understanding of and proactive measures taken to address the present state of India's forests and fisheries are imperative for ensuring their sustainable management and conservation, thereby safeguarding these invaluable ecosystems for future generations.

Encompassing a vast expanse of 4.06 billion ha (ha), constituting approximately 31% of the world's total land area, are the sprawling forests that embellish our planet. It is essential to acknowledge that these awe-inspiring woodland expanses are not evenly distributed across the globe, but rather exhibit an uneven distribution. Nevertheless, this extensive coverage of forests equates to an estimated allocation of approximately 0.52 ha per person. Amidst this resplendent tapestry, the tropical domain commands the largest share, accounting for 45% of the world's forests, closely followed by the boreal (21%), temperate (16%), and subtropical (11%) regions. Remarkably, a mere five nations collectively serve as custodians of more than half (54%) of the world's forests, with the Russian Federation leading the way (20%), trailed by Brazil (12%), Canada (9%), the United States of America (8%), and China (5%).

Since 1990, 178 million acres of forest have been destroyed worldwide. From 7.8 million ha per year in the decade 1990–2000 to 5.2 million ha per year in 2000–2010 and 4.7 million ha per year in 2010–2020, the rate of net forest loss decreased. Due to a reduction in the rate of forest expansion, the rate of decline of net forest loss slowed in the most recent ten years.

Between 2010 and 2020, Africa experienced the highest yearly rate of net forest loss (3.9 million ha), followed by South America (2.6 million ha). In each of the three decades after 1990, the rate of net forest loss in Africa has accelerated. However, compared to 2000–2010, it has significantly decreased in South America, with the rate there hovering around 50% in 2010–2020. Asia, Oceania, and Europe saw the largest net gains in forest area between 2010 and 2020. However, the net gain rates in 2010–2020 were significantly lower in both Europe and Asia than they were in 2000–2010. Oceania saw net forest area decreases during the years 1990 and 2000 and 2000 and 2010.

Although the rate of forest loss has significantly decreased, an estimated 420 million acres of forest have been lost worldwide through deforestation since 1990. Deforestation was expected to have decreased from 12 million ha in 2010–2015 to 10 million ha in the most recent five-year period (2015–2020).

The majority of the world's forests (3.75 billion ha) consist of naturally regenerating forests, whereas 7% (290 million ha) are planted. Since 1990, the area of naturally regenerating forests has shrunk (at a decreasing rate of loss), while the number of planted forests has grown by 123 million ha. In the past ten years, the area of planted forests has increased at a slower rate. India's forest cover has grown over time. It has increased from 640,819 square kilometres (sq. km) in 1987 to 701,673 sq. km in 2015 (Table 1.4 and Figure 1.3).

The amount of actual woodland has grown slowly. Actual forest cover in 1995 was 638,879 sq. km; in 2019, it was 71,224. It has increased by 11.49% over the past 23 years (77,370 sq. km). While protected areas have seen a decline, both reserved and unclassified areas have seen an increase. The changes observed in the forest cover (in sq. km) can be attributed to factors such as deforestation, afforestation efforts, land-use changes, climate change

Table 1.4 India's Forest Cover (1987–2015)

Year	Forest cover (in sq. km)
1987	640,819
1989	638,804
1991	639,364
1993	639,386
1995	638,879
1997	633,397
1999	637,293
2001	653,898
2003	677,816
2005	690,171
2009	692,394
2011	690,899
2013	697,898
2015	701,673

Source: Ministry of Agriculture & Farmers Welfare, Govt. of India (ON3410 & past issues). Data Source: indiastat.com. Data accessed on 25 July 2023.

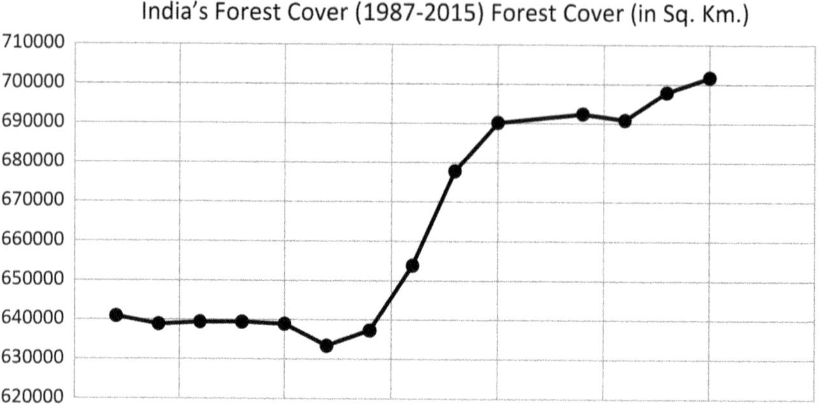

Figure 1.3 India's Forest Cover (1987–2015).

impacts, and natural disturbances like wildfires or disease outbreaks. These changes have significant implications for biodiversity loss, carbon emissions, soil erosion, water regulation, and overall ecological balance, affecting wildlife habitats, climate resilience, and the livelihoods of communities dependent on forests. In 2004, there were 690,171 sq. km of forest, of which 690,171 sq. km were open forests,12% of which were fairly dense, and 1.66% of which were heavily forested (54,569 sq. km). The proportion of very dense areas, which made up 3.02% of all geographical areas and covered 99,278 sq. km, moderately dense areas, which made up 9.38% of all geographical areas (308,472 sq. km), and open forest, which made up 9.26% of all geographical areas, increased to 712,249 sq. km in 2017–2018 (Table 1.5).

Table 1.5 Area under Forest by Type in India (in sq. km.)

Year	Geographical area	Actual forest cover	Very dense forest	Moderately dense forest,	Open forest	Mangrove	Scrub	Non-forest
1995	3,287,263	765,210	385,037	–	249,309	4,533	60,528	2,587,856
1997	3,287,263	765,210	377,358	–	255,064	4,837	57,211	2,596,655
1999	3,287,263	765,253	377,358	–	255,064	4,871	51,896	2,598,074
2001	3,287,263	653,898	395,169		258,729	4,482	47,318	2,586,047
2003	3,287,263	677,816	54,518	334,056	289,242	4,448	40,269	2,569,178
2005	3,287,263	690,171	83,472	319,948	286,751	4,639	38,475	2,558,617
2009	3,287,263	690,899	83,510	319,012	288,377	4,581	41,525	2,554,839
2011	3,287,263	692,027	83,471	320,736	287,820	4,663	42,176	2,553,060
2013	3,287,263	697,898	83,502	318,745	295,651	4,628	41,383	2,547,982
2015	3,287,263	701,673	85,904	315,374	300,395	4,740	41,362	2,544,228
2017	3,287,469	708,273	98,158	308,318	301,797	4,921	45,979	2,533,217

Source: Ministry of Agriculture & Farmers Welfare, Govt. of India (ON3410 & past issues). Data Source: indiastat.com. Data accessed on 25 July 2023.

Rajasthan emerges as the largest state geographically, covering 342,239 sq. km, followed closely by Madhya Pradesh with 308,252 sq. km. In contrast, states like Goa and Sikkim have very small geographical areas of 3,702 and 7,096 sq. km. Andaman & Nicobar Islands, Dadra and Nagar Haveli, and Daman and Diu are the largest UTs geographically with 8,249 sq. km and 602 sq. km. Lakshadweep and Chandigarh have very small areas of 30 and 114 sq. km, respectively. States like Arunachal Pradesh, Manipur, and Mizoram, despite their relatively smaller populations, have substantial geographical areas 83,743, 22,327, and 21,081 sq. km, respectively. Madhya Pradesh stands out with the largest area of reserved forests at 61,886 sq. km, reflecting its commitment to conservation. States like Maharashtra (50,865 sq. km), and Odisha (36,049 sq. km) also maintain substantial reserved forest areas, crucial for biodiversity preservation and mitigating climate change impacts.

Protected forests, which allow sustainable use under regulated conditions, show notable differences across states. Madhya Pradesh leads in protected forests with 31,098 sq. km, followed by Himachal Pradesh (28,887 sq. km) and Chhattisgarh (24,036 sq. km). These forests play a vital role in supporting local livelihoods through sustainable harvesting practices while safeguarding biodiversity. Unclassed forests, areas not designated under specific protection categories, indicate variations in forest management approaches. States like Arunachal Pradesh (27,312 sq. km), Manipur 13,180 (sq. km), and Chhattisgarh (9,883 sq. km) have significant unclassed forest areas, highlighting potential areas for future classification and conservation efforts. Andaman & Nicobar Islands stands out with the highest recorded forest area percentage at 86.93%, reflecting its significant dedication to preserving large swathes of land for conservation purposes. This is followed by Sikkim (82.31%) and Manipur (78.01%), indicating substantial efforts in maintaining protected forest areas to safeguard biodiversity and ecosystem services.

On the other end of the spectrum, states like Delhi and Lakshadweep have negligible or no designated reserved forest areas, highlighting the challenges in balancing urban development and conservation priorities. States with moderate recorded forest area percentages, such as Arunachal Pradesh (61.55%) and Tripura (60.02%), show a balanced approach in managing forest resources for sustainable use and conservation. These areas often employ sustainable forestry practices that promote biodiversity conservation while allowing for regulated economic activities (Table 1.6).

The forest ecosystem has the ability to absorb carbon from the atmosphere and store it, primarily in living biomass and soil but also, to a lesser extent, in dead wood and litter. Table 1.7 provides a detailed overview of carbon stocks across different forest types in India as of 2021, focusing on various carbon pools such as above-ground biomass (AGB), below-ground biomass (BGB), Dead Wood, Litter, and soil organic carbon (SOC), measured in million metric tons (Mt). The total carbon stock in India's forests

Table 1.6 State-wise Recorded Forest Area in India in '000 sq. km

States/UTs	Geographical Area	Recorded forest area (in different categories)			Total RFA	%age of Geographical area
		Reserved forests	Protected forests[2]	Unclassed forests		
Andaman & Nicobar Islands	8,249	5,613	1,558	0	7,171	86.93
Andhra Pradesh	162,968	31,959	5,069	230	37,258	22.86
Arunachal Pradesh	83,743	12,371	11,857	27,312	51,540	61.55
Assam	78,438	17,864	0	8,972	26,836	34.21
Bihar	94,163	693	6,183	566	7,442	7.90
Chandigarh	114	32	0	3	35	30.70
Chhattisgarh	135,192,	25,897	24,036	9,883	59,816	44.25
Dadra and Nagar Haveli and Daman and Diu	602	203	5	6	214	35.55
Delhi	1,483	78	25	0	103	6.95
Goa	3,702	119	755	397	1,271	34.33
Gujarat	196,244	14,574	2,898	4,398	21,870	11.14
Haryana	44,212	249	1,158	152	1,559	3.53
Himachal Pradesh	55,673	1,883	28,887	7,178	37,948	68.16
Jammu & Kashmir	54,624	17,648	2,551	0	20,199	36.98
Jharkhand	79,716	4,500	18,922	1,696	25,118	31.51
Karnataka	191,791	28,690	3,931	5,663	38,284	19.96
Kerala	38,852	11,522	0	0	11,522	29.66
Ladakh	168,055	7	0	0	7	0.00
Lakshadweep	30	0	0	0	0	0.00
Madhya Pradesh	308,252	61,886	31,098	1,705	94,689	30.72
Maharashtra	307,713	50,865	6,433	4,654	61,952	20.13
Manipur	22,327	984	3,254	13,180	17,418	78.01
Meghalaya	22,429	1,113	12	8,371	9,496	42.34
Mizoram	21,081	4,499	1,823	1,157	7,479	35.48
Nagaland	16,579	234	0	8,389	8,623	52.01
Odisha	155,707	36,049	25,133	22	61,204	39.31
Puducherry	49	0	2	11	13	26.53
Punjab	50,362	44	1,137	1,903	3,084	6.12
Rajasthan	342,239	12,176	18,543	2,144	32,863	9.60
Sikkim	7,096	5,452	389	0	5,841	82.31
Tamil Nadu	130,060	20,523	1,053	1,612	23,188	17.83
Telangana	112,077	25,800	1,592	296	27,688	24.70
Tripura	10,486	3,588	2	2,704	6,294	60.02
Uttar Pradesh[1]	240,928	11,560	296	5,528	17,384	7.22
Uttarakhand	53,483	26,547	9,885	1,568	38,000	71.05
West Bengal	88,752	7,054	3,772	1,053	11,879	13.38
India	**3,287,469**	**442,276**	**212,259**	**120,753**	**775,288**	**23.58**

Source: Ministry of Agriculture & Farmers Welfare, Govt. of India. (ON3410 & past issues).
Data source: indiastat.com. Data accessed on 25 July 2023.

Table 1.7 Carbon Stock in Different Forest Types under Different Pools in India (2021) (in '000 Tonne)

Forest type stratum	Area (in sq. km)	Above-ground biomass (AGB)	Below-ground biomass (BGB)	Dead wood	Litter	Soil organic carbon (SOC)	Total
Tropical Wet Evergreen Forests	19,572	146,678	54,269	4,859	5,351	134,456	345,613
Tropical Semi-Evergreen Forests	69,195	242,270	53,306	4,992	13,092	372,346	686,006
Tropical Moist Deciduous Forests	131,805	422,838	93,027	7,955	28,040	750,821	1,302,681
Littoral and Swamp Forest	5,478	28,046	10,377	273	928	32,999	72,623
Tropical Dry Deciduous Forests	280,547	607,145	254,946	9,107	29,121	1,276,497	2,176,816
Tropical Thorn Forests	13,259	11,214	4,709	283	904	32,516	49,626
Tropical Dry Evergreen Forests	835	2,763	1,161	40	89	3,696	7,749
Subtropical Broadleaved Hill Forests	31,015	111,853	46,978	2,057	4,870	266,818	432,576
Subtropical Pine Forests	17,801	87,843	23,715	1,870	2,788	123,141	239,357
Subtropical Dry Evergreen Forest	173	1,137	478	5	8	1,070	2,698
Montane Wet Temperate Forests	20,185	84,471	22,804	3,387	2,227	229,584	342,473
Himalayan Moist Temperate Forests	28,727	318,037	85,867	7,753	8,431	226,630	646,718
Himalayan Dry Temperate Forests	4,255	51,927	15,059	583	1,730	34,601	103,900
Sub-alpine Forests	12,672	89,109	25,837	2,477	2,543	112,476	232,442
Most Alpine Scrub	652	1,445	420	25	57	3,687	5,634
Dry Alpine Scrub	2,396	13,856	4,019	227	315	9,041	27,458
Plantation/ TOF	75,221	99,277	21,878	1,772	6,758	399,790	529,475
Total	713,789	2,319,909 (32.50)	718,850 (10.70)	47,665 (0.67)	107,252 (1.50)	4,010,169 (56.18)	7,203,845 (100.92)

Source: Ministry of Agriculture & Farmers Welfare, Govt. of India (ON3410 & past issues).

Data Source: indiastat.com. Data accessed on 25 July 2023.

is estimated at approximately 7,203.8 Mt, with 56.18% of total carbon sink from SOC with 4,010.2 Mt, 32.50% of total carbon sink from AGB with 2,319.9 Mt, 10.7% of total carbon stock from BGB with 718.5 Mt, 1.5% of total carbon stock from litters with 107.25 Mt, and 0.67% of total carbon stock from dead wood with 47.65 Mt. Among the forest types, Tropical Dry Deciduous Forests exhibit the highest total carbon stock at 2,176.8 Mt, emphasizing their role as substantial carbon sinks due to their extensive biomass and organic carbon storage. Himalayan Moist Temperate Forests and Subtropical Broadleaved Hill Forests also show notable carbon stocks, highlighting their ecological importance in carbon sequestration. This substantial storage is primarily attributed to their extensive AGB and substantial SOC content. These forests play a crucial role in mitigating climate change by sequestering carbon dioxide from the atmosphere through their dense vegetation cover and deep-rooted organic matter. Conversely, Most Alpine Scrub forests exhibit the lowest carbon stocks at 5,634,000 tonnes, reflecting their sparse vegetation and limited biomass. The stark contrast between these forest types underscores the importance of ecosystem diversity in carbon sequestration efforts. Above-ground biomass consistently emerges as the predominant carbon pool across all forest types, highlighting the critical role of tree growth and biomass accumulation in carbon storage. In contrast, soil organic carbon varies significantly across different forest types, influenced by factors such as temperature, moisture, and vegetation characteristics.

Table 1.8 Arunachal Pradesh stands out with the highest total carbon stock of 1,023.8 Mt, driven largely by its substantial SOC (560.3 Mt) and AGB (340.4 Mt). Other states with significant carbon stocks include Madhya Pradesh (609.3 Mt), Maharashtra (451.6 Mt), and Chhattisgarh (496.4 Mt). Conversely, smaller states like Delhi and Puducherry exhibit lower total carbon stocks, reflecting their limited forest cover and biomass. States with dense forest cover and diverse ecosystems, like Arunachal Pradesh and Madhya Pradesh, play a crucial role in mitigating climate change through carbon storage.

Table 1.9 provides a comprehensive overview of tree cover across different states and UTs in India for the year 2021, measured in sq. km. Andaman & Nicobar Islands, with a geographical area of 8,249 sq. km, has a minimal tree cover of 23 sq. km, translating to a mere 0.28% of its land area. Andhra Pradesh, spanning 162,968 sq. km, boasts a significantly higher tree cover of 4,679 sq. km, representing 2.87% of its total area. Arunachal Pradesh and Assam, with geographic areas of 83,743 sq. km and 78,438 sq. km, respectively, show tree cover percentages of 1.20% and 2.08%. Bihar, covering 94,163 sq. km, has 2,341 sq. km of tree cover, accounting for 2.49% of its area. Notably, Chandigarh, despite its small geographical area of 114 sq. km, has a high tree cover percentage of 13.16%, albeit with only 15 sq. km of tree cover. Chhattisgarh, with 135,192 sq. km, has 5,355 sq. km under tree cover,

Table 1.8 State-wise Forest Carbon Stock under Different Carbon Pools in India (2021) (in '000 Tonne)

States/UTs	Area (in sq. km)	Above-ground biomass (AGB)	Below-ground biomass (BGB)	Dead wood	Litter	Soil organic carbon (SOC)	Total
Andaman & Nicobar Islands	6,744	47,560	15,450	1,432	1,808	43,586	109,836
Andhra Pradesh	29,784	63,951	25,064	979	3,171	137,057	230,222
Arunachal Pradesh	66,431	340,351	102,229	9,163	11,802	560,298	1,023,843
Assam	28,312	87,070	21,495	1,875	4,890	156,042	271,372
Bihar	7,381	14,743	5,249	231	785	35,873	56,881
Chandigarh	23	47	15	1	3	117	183
Chhattisgarh	55,717	152,714	48,947	2,520	8,487	283,769	496,437
Delhi	195	263	78	5	17	839	1,202
Goa	2,244	8,863	2,606	232	448	13,095	25,244
Gujarat	14,926	28,602	9,814	502	1,634	67,214	107,766
Haryana	1,603	2,326	836	41	139	6,890	10,232
Himachal Pradesh	15,443	114,269	31,880	2,657	3,328	105,937	258,071
Jammu & Kashmir	21,387	163,897	45,864	3,386	4,951	152,772	370,870
Jharkhand	23,721	51,017	20,819	774	2,536	109,665	184,811
Karnataka	38,730	122,741	36,716	2,890	6,380	207,668	376,395
Kerala	21,253	61,802	17,440	1,534	3,198	121,549	205,523
Ladakh	2,272	13,293	3,836	269	317	12,897	30,702
Madhya Pradesh	77,493	171,587	67,160	2,676	8,653	359,174	609,250
Maharashtra	50,798	137,831	42,353	2,316	7,928	261,178	451,606
Manipur	16,598	47,590	14,101	880	2,652	111,708	176,931
Meghalaya	17,046	55,241	15,820	1,238	3,075	108,014	183,388
Mizoram	17,820	48,157	10,622	758	3,140	95,961	158,638
Nagaland	12,251	39,339	10,618	854	2,006	82,115	134,932
Odisha	52,156	131,015	40,441	2,252	7,671	263,451	444,830
Puducherry	53	76	17	1	5	287	386
Punjab	1,847	3,420	1,284	56	175	8,623	13,558
Rajasthan	16,655	26,714	10,803	462	1,476	71,319	110,774
Sikkim	3,341	18,024	5,466	498	607	30,944	55,539
Tamil Nadu	26,419	60,459	20,671	1,198	3,102	129,183	214,613
Telangana	21,214	44,413	18,415	675	2,169	96,314	161,986
Tripura	7,722	24,349	5,358	477	1,486	43,304	74,974
Uttar Pradesh	14,818	32,543	10,234	534	1,825	72,105	117,241
Uttarakhand	24,305	159,674	42,893	3,561	5,184	166,847	378,159
West Bengal	16,832	45,365	14,119	726	2,162	92,889	155,261
India	713,789	2,319,910	718,852	47,665	107,251	4,010,168	7,203,846
Total Percentage		(32.50)	(10.07)	(0.67)	(1.50)	(56.18)	(100.92)

Source: Ministry of Environment & Forests, Govt. of India (ON2854).
Data Source: indiastat.com. Data accessed on 25 July 2023.

Table 1.9 State-wise Estimates of Tree Cover in India (2021) (in sq. km)

States/UTs	Geographic area	Tree cover	%age of tree cover to geographical area
Andaman & Nicobar Islands	8,249	23	0.28
Andhra Pradesh	162,968	4,679	2.87
Arunachal Pradesh	83,743	1,001	1.20
Assam	78,438	1,630	2.08
Bihar	94,163	2,341	2.49
Chandigarh	114	15	13.16
Chhattisgarh	135,192	5,355	3.96
Dadra and Nagar Haveli and Daman and Diu	602	32	5.32
Delhi	1,483	147	9.91
Goa	3,702	244	6.59
Gujarat	196,244	5,489	2.80
Haryana	44,212	1,425	3.22
Himachal Pradesh	55,673	675	1.21
Jammu & Kashmir	54,624	3,511	6.43
Jharkhand	79,716	2,867	3.60
Karnataka	191,791	7,494	3.91
Kerala	38,852	2,820	7.26
Ladakh	168,055	954	0.57
Lakshadweep	30	0	0.17
Madhya Pradesh	308,252	8,054	2.61
Maharashtra	307,713	12,108	3.93
Manipur	22,327	169	0.76
Meghalaya	22,429	698	3.11
Mizoram	21,081	444	2.11
Nagaland	16,579	365	2.20
Odisha	155,707	5,004	3.21
Puducherry	490	23	4.69
Punjab	50,362	1,138	2.26
Rajasthan	342,239	8,733	2.55
Sikkim	7,096	39	0.55
Tamil Nadu	130,060	4,424	3.40
Telangana	112,077	2,848	2.54
Tripura	10,486	228	2.17
Uttar Pradesh	240,928	7,421	3.08
Uttarakhand	53,483	1,001	1.87
West Bengal	88,752	2,349	2.65
India	**3,287,469**	**95,748**	**2.91**

Source: Ministry of Environment & Forests, Govt. of India (ON2854).
Data Source: indiastat.com. Data accessed on 25 July 2023.

equating to 3.96%. The combined entity of Dadra and Nagar Haveli and Daman and Diu, with a geographical area of 602 sq. km, shows 32 sq. km of tree cover, or 5.32% of its land. Delhi, with 1,483 sq. km, has 147 sq. km of tree cover, resulting in a 9.91% coverage. Goa, with 3,702 sq. km, and Gujarat, with 196,244 sq. km, report tree cover percentages of 6.59% and 2.80%, respectively. Haryana's tree cover spans 1,425 sq. km of its 44,212 sq. km area, yielding a 3.22% coverage. Himachal Pradesh and Jammu & Kashmir, with geographic areas of 55,673 sq. km and 54,624 sq. km, have tree cover percentages of 1.21% and 6.43%, respectively. Jharkhand's tree cover of 2,867 sq. km makes up 3.60% of its 79,716 sq. km area, while Karnataka, with a large geographic area of 191,791 sq. km, has 7,494 sq. km under tree cover, or 3.91%. Kerala's significant tree cover of 2,820 sq. km constitutes 7.26% of its 38,852 sq. km area. Ladakh and Lakshadweep, with geographic areas of 168,055 sq. km and 30 sq. km, show tree cover percentages of 0.57% and 0.17%, respectively. Madhya Pradesh, covering 308,252 sq. km, has a substantial tree cover of 8,054 sq. km, making up 2.61% of its area. Maharashtra, similarly expansive at 307,713 sq. km, reports a high tree cover of 12,108 sq. km, or 3.93%. Manipur and Meghalaya, with geographic areas of 22,327 sq. km and 22,429 sq. km, have tree cover percentages of 0.76% and 3.11%, respectively. Mizoram and Nagaland, with areas of 21,081 sq. km and 16,579 sq. km, report tree cover percentages of 2.11% and 2.20%. Odisha's tree cover spans 5,004 sq. km of its 155,707 sq. km area, resulting in a 3.21% coverage. Puducherry, Punjab, and Rajasthan, with geographic areas of 490 sq. km, 50,362 sq. km, and 342,239 sq. km, show tree cover percentages of 4.69%, 2.26%, and 2.55%, respectively. Sikkim, with a small geographic area of 7,096 sq. km, has a tree cover percentage of 0.55%. Tamil Nadu, Telangana, and Tripura, with areas of 130,060 sq. km, 112,077 sq. km, and 10,486 sq. km, report tree cover percentages of 3.40%, 2.54%, and 2.17%. Uttar Pradesh and Uttarakhand, with geographic areas of 240,928 sq. km and 53,483 sq. km, have tree cover percentages of 3.08% and 1.87%, respectively. West Bengal, covering 88,752 sq. km, has 2,349 sq. km of tree cover, equating to 2.65% of its area. Overall, India, with a total geographic area of 3,287,469 sq. km, has a tree cover of 95,748 sq. km, representing 2.91% of its land area.

The fisheries industry has been acknowledged as a significant source of income and employment because it fosters the expansion of several related industries, provides access to low-cost, nutrient-rich food, and provides a means of subsistence for a sizable portion of the nation's economically disadvantaged population. The socioeconomic development of the nation's fisheries sector is significant. Over 28 million people in India are employed and receive income from the country's rapidly expanding sector of fishing, which also supplies nutrition and food security to a huge portion of the population. The "Sunrise Sector" designation has been given to the fishing industry, which has had excellent double-digit average annual growth of 10.87% since 2014–2015. The sector has enormous growth potential and produced a record-breaking 142 lakh tons of fish in FY 2019–2020. Additionally, it has helped to support the livelihoods of more than 28 million people in India, particularly in marginalized and vulnerable groups, and it has aided in promoting socioeconomic growth.

Table 1.10 Production of Fish in India ('000 Tonne)

Year	Gross	Waste	Net
1951	731	74	677
1956	1,012	99	913
1961	961	94	867
1966	1,367	134	1,233
1971	1,852	107	1,745
1976	2,174	100	2,074
1981	2,443	183	2,260
1986	2,921	169	2,752
1991	4,044	381	3,663
1996	5,258	336	4,922
1997	5,380	345	5,035
1998	5,319	323	4,921
1999	5,586	347	5,239

Source: Ministry of Environment & Forests, Govt. of India (ON2854).

Data Source: indiastat.com. Data accessed on 25 July 2023.

India, the world's second-largest producer of fish, accounts for 7.56% of global production, 1.24% of the country's GVA, and more than 7.28% of its agricultural GVA. Fishing and aquaculture continue to be important sources of food, nutrition, income, and a way of life for millions of people. The Fisheries sector produced export earnings of Rs. 46,662.85 crores in 2019–2020. The sector provides livelihood assistance to around 280 lakh individuals at the primary level, and approximately twice that many persons farther down the value chain. The fishing business has grown at a 7% yearly rate during the last few years. Because it is affordable and abundant, fish is one of the healthiest alternatives to animal protein (see https://www.pib.gov.in/PressReleasePage.aspx?PRID=1786303).

India's gross fish production climbed from 731,000 tonnes to 5,586,000 tonnes between 1951 and 1999, a rise of 664.16%. Additionally, there has been a 673.86% increase in net fish production and a 78.68% increase in trash. India's net fish production in 1951 was 677,000 tonnes; by the end of 1999, it had climbed to 523,900 tonnes. Similarly, waste production in 1951 was 74,000 tonnes; by the end of 1999, it had decreased to 347,000 tonnes (Table 1.10).

In 1950–1951, India produced 752,000 tonnes of fish, of which 534,000 tonnes came from the sea and 218,000 tonnes from the inland. By the end of 2019–2020, however, this fish production will have climbed to 14,164,000 tonnes, comprising 10,437,000 tonnes of inland fish and 3,727,000 tonnes of marine fish (Table 1.11).

Table 1.11 Production of Fish in India

Year	Fish production (in `000 Tonne)			Average annual growth rate (in %)		
	Marine	Inland	Total	Marine	Inland	Total
1950–1951	534	218	752	–	–	–
1955–1956	596	243	839	2.32	2.29	2.31
1960–1961	880	280	1,160	9.53	3.05	7.65
1965–1966	824	507	1,331	–1.27	16.21	2.95
1970–1971	1,086	670	1,756	6.36	6.43	6.39
1973–1974	1,210	748	1,958	3.81	3.88	3.83
1978–1979	1,490	816	2,306	4.63	1.82	3.55
1979–1980	1,492	848	2,340	0.13	3.92	1.47
1980–1981	1,555	887	2,442	4.22	4.60	4.36
1981–1982	1,445	999	2,444	–7.07	12.63	0.08
1982–1983	1,427	940	2,367	–1.25	–5.91	–3.15
1983–1984	1,519	987	2,506	6.45	5.00	5.87
1984–1985	1,698	1,103	2,801	11.78	11.75	11.77
1985–1986	1,716	1,160	2,876	1.06	5.17	2.68
1986–1987	1,713	1,229	2,942	–0.17	5.95	2.29
1987–1988	1,658	1,301	2,959	–3.21	5.86	0.58
1988–1989	1,817	1,335	3,152	9.59	2.61	6.52
1989–1990	2,275	1,402	3,677	25.21	5.82	16.66
1990–1991	2,300	1,536	3,836	1.10	9.56	4.32
1991–1992	2,447	1,710	4,157	6.39	11.33	8.37
1992–1993	2,576	1,789	4,365	5.27	4.62	5.00
1993–1994	2,649	1,995	4,644	2.83	11.51	6.39
1994–1995	2,692	2,097	4,789	1.62	5.11	3.12
1995–1996	2,707	2,242	4,949	0.56	6.91	3.34
1996–1997	2,967	2,381	5,348	9.60	6.20	8.06
1997–1998	2,950	2,438	5,388	–0.57	2.39	0.75
1998–1999	2,696	2,602	5,298	–8.61	6.73	1.67
1999–2000	2,852	2,823	5,675	5.79	8.49	7.12
2000–2001	2,811	2,845	5,656	–1.44	0.78	–0.33
2001–2002	2,830	3,126	5,956	0.68	9.88	5.30
2002–2003	2,990	3,210	6,200	5.65	2.69	4.10
2003–2004	2,941	3,458	6,399	–1.64	7.73	3.21
2004–2005	2,779	3,526	6,305	–5.51	1.97	–1.47
2005–2006	2,816	3,756	6,572	1.33	6.52	4.23
2006–2007	3,024	3,845	6,869	7.39	2.37	4.52
2007–2008	2,920	4,207	7,127	–3.44	9.41	3.76
2008–2009	2,978	4,638	7,616	1.99	10.24	6.86
2009–2010	3,104	4,894	7,998	4.23	5.52	5.02
2010–2011	3,250	4,981	8,231	4.70	1.78	2.91
2011–2012	3,372	5,294	8,666	3.75	6.28	5.28
2012–2013	3,321	5,719	9,040	–1.51	8.03	4.32
2013–2014	3,440	6,132	9,576	3.67	7.29	5.96
2014–2015	3,502	6,929	10,431	3,569	6,691	10,260
2015–2016	3,583	7,213	10,795	3,600	7,162	10,762
2016–2017	3,625	7,806	11,431	3,625	7,806	11,431
2017–2018	3,756	8,948	12,704	3,688	8,902	12,590
2018–2019	3,853	9,720	13,573	3,853	9,720	13,573
2019–2020	3,727	10,437	14,164	3,727	10,437	14,164

Source: Ministry of Agriculture & Farmers Welfare, Govt. of India (ON2738 & past issues). Data Source: indiastat.com. Data accessed on 25 July 2023.

1.5   Renewable Energy

India has steadfastly pursued the development and utilization of renewable energy sources to cater to its burgeoning energy demands, all while earnestly addressing pressing environmental concerns. Leveraging its expansive geographical diversity and abundant natural resources, the nation possesses immense potential for generating renewable energy.

Renewable energy in India encompasses solar, wind, biomass, and hydropower. Solar energy has emerged as a prominent and rapidly expanding source, capitalizing on India's ample sunlight throughout the year. The country has achieved significant milestones in solar power generation, marked by the establishment of large-scale solar parks and the growing prevalence of rooftop installations. The government has implemented policies, initiatives, and mechanisms to foster the adoption of solar energy, including the provision of financial incentives, subsidies, and targeted programmes.

Wind power also assumes a pivotal role as a renewable energy source in India. The nation boasts favourable wind conditions in various regions, resulting in the establishment of numerous wind farms. Strategically positioned wind turbines across the landscape make substantial contributions to the electricity grid, thereby reducing reliance on conventional fossil fuel-based power generation.

Biomass energy, derived from organic matter such as agricultural waste and crop residues, constitutes a noteworthy component of India's renewable energy portfolio. Biomass power plants not only generate electricity but also offer a sustainable solution for waste management and the effective disposal of agricultural residues.

Hydropower, harnessed from the energy potential of flowing or falling water, has been a stalwart in India for several decades. Large-scale hydropower projects tap into the energy resources of rivers, making substantial contributions to the national grid. Additionally, small-scale micro-hydropower systems are deployed in remote areas to meet localized energy requirements.

The Indian government has set ambitious targets to amplify the proportion of renewable energy in the overall energy mix. Endeavours such as the National Solar Mission and the National Wind Mission aim to expedite the expansion of renewable energy infrastructure and attract investments to the sector. Simultaneously, India actively fosters research and development in renewable energy technologies, exploring innovative solutions such as offshore wind farms and floating solar power plants.

The transition towards renewable energy in India not only facilitates the reduction of greenhouse gas emissions and the mitigation of climate change but also presents ample opportunities for sustainable economic growth, employment generation, and enhanced energy access in remote and rural areas. The renewable energy sector has witnessed substantial domestic and international investments, nurtured a dynamic market and fostered an environment conducive to technological advancements and collaborative initiatives.

India's commitment to renewable energy underscores its vision of a clean and sustainable future. With a holistic approach that encompasses policy support, technological innovation, and stakeholder engagement, the country aims to tap into its renewable energy potential and emerge as a global leader in the transition to a low-carbon economy.

The potential for renewable energy in India expanded by 250% between 2014 and 2021, according to Prime Minister Mr. Narendra Modi's statement in June 2021. India's electricity needs, according to the Central Electricity Authority, are expected to increase and reach 817 GW by 2030. The real estate and transportation sectors will generate the majority of the demand. India's electricity needs, according to the Central Electricity Authority, are expected to increase and reach 817 GW by 2030. The real estate and transportation sectors will generate the majority of the demand.

India is fifth in the world for solar power capacity, fourth for wind power, and fourth for renewable energy capacity. India maintained its third-place position on the EY Renewable Energy Country Attractive Index 2021 in October 2021. The government provided Rs. 19,500 crore ($2.57 billion) in the Union Budget 2022–2023 for a PLI scheme to promote the production of high-efficiency solar modules. India started the Mission Creation Cleantech Exchange, a global programme that will hasten the innovation of clean energy (https://www.ibef.org/industry/renewable-energy).

India has produced 294,106.14 million unit of electricity from renewable energy sources in 2019–2020, which is increased by 1.17% from 294,106.14 million units to 297,547.03 million units in the next year. In 2021–2022, it is increased by 8.40%, i.e. 24,992.6 million units from its previous level (Table 1.12 and Figure 1.4).

India was holding 71,240,000 tonnes of coal, 18,505,000 tonnes of petroleum, and 647 million cubic metres of natural gas, which has the power of producing 276,666 Gwh of electricity. Over the time India has discovered its coal reserves. By the end of 2020–2021, India holds 955,054,000 tonnes of coal, 35,876,000 of lignite, 228,605,000 crude petroleum and 60,645 million

Table 1.12 Electricity Generated from Renewable Energy Sources in India (2019–2020 to 2022–2023-up to May 2022) (in Million Unit)

Year	*RE generation (including large hydro)*
2019–2020	294,106.14
2020–2021	297,547.03
2021–2022	322,539.63
2022–2023-up to May 2022	60,411.70

Source: Lok Sabha Unstarred Question No. 1945, dated on 28 July 2022. Data Source: indiastat.com. Data accessed on 25 July 2023.

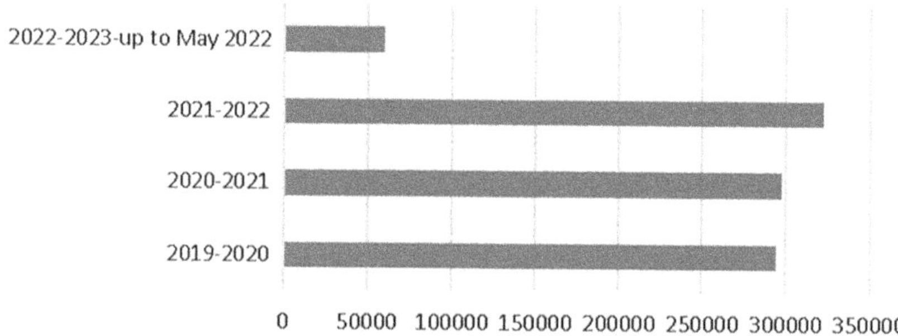

Figure 1.4 Electricity Generated from Renewable Energy Sources in India (2019–2020 to 2022–2023-up to May 2022).

cubic metres of natural gas, which have the power to produce 1,315,110 GWh of electricity (Table 1.13).

Consumption of conventional energy has been the primary source of electricity production in India, in which coal plays an important role. In 1970–1971, India used 71,230,000 tonnes of coal, 3,390,000 tonnes of lignite, and 18,379,000 tonnes of crude petroleum, 647 million Cubic Metres of natural gas in order to produce 43,724 GWh of electricity. With the increased production of electricity over time, the consumption of coal, lignite, crude petroleum, and natural gas has also increased. By the end of 2020–2021, India has produced 1,237,800 GWh of electricity by consuming 905,883,000 tonnes of coal, 221,773,000 tonnes of crude petroleum, and 60,645 million cubic metres of natural gas. Over 51 years, India's coal consumption has increased by over 1,171.78%, i.e., from 71,230,000 in 1970–1971 to 905,883,000 in 2020–2021 (Table 1.14).

1.6 Coal and Minerals

India possesses abundant natural resources, including substantial reserves of coal and minerals, which play a pivotal role in the nation's economic development and industrial progress. Renowned globally, India's coal and mineral reserves rank among the largest, positioning the country as a prominent producer and consumer of these invaluable resources.

The coal reserves in India are primarily concentrated in states such as Jharkhand, Odisha, Chhattisgarh, West Bengal, and Madhya Pradesh. These reserves encompass significant quantities of both metallurgical coal, vital for the steel industry, and thermal coal, crucial for power generation. The coal mining sector has been a cornerstone of India's economy for many years, catering to diverse sectors and ensuring energy security.

Table 1.13 Availability of Primary Sources of Conventional Energy in India (1970–1971 to 2020–2021)

(in ' 000 tonne)

Year	Coal	Lignite	Crude petroleum	Natural gas (in million cubic metre)	Electricity[1] hydro & nuclear (in GWh)
1970–1971	71,240	–	18,505	647	27,666
1971–1972	74,070	–	20,250	718	29,214
1972–1973	78,210	–	19,405	771	28,329
1973–1974	77,680	–	21,044	762	31,368
1974–1975	85,600	–	21,700	951	30,081
1975–1976	92,170	–	22,072	1,124	35,928
1976–1977	100,100	–	22,946	1,381	38,088
1977–1978	103,840	–	25,270	1,464	40,279
1978–1979	100,160	–	26,290	1,711	49,929
1979–1980	105,540	–	27,887	1,676	48,354
1980–1981	109,320	–	26,755	1,522	49,543
1981–1982	121,020	–	30,654	2,222	52,586
1982–1983	130,138	–	33,460	2,957	50,396
1983–1984	139,302	–	36,465	3,401	53,500
1984–1985	141,458	–	36,154	4,141	58,023
1985–1986	155,536	–	44,784	4,950	56,003
1986–1987	166,846	–	45,956	7,075	58,862
1987–1988	180,645	–	48,091	7,968	52,479
1988–1989	193,970	–	49,855	9,250	63,690
1989–1990	209,481	–	53,577	11,172	66,741
1990–1991	214,987	–	53,720	12,765	77,782
1991–1992	232,338	–	54,340	14,441	78,282
1992–1993	245,255	–	56,197	16,116	76,595
1993–1994	256,327	–	57,848	16,340	75,861
1994–1995	269,180	–	59,588	17,337	88,360
1995–1996	284,043	–	62,509	20,932	80,561
1996–1997	298,621	–	66,806	21,324	77,972
1997–1998	306,824	–	68,351	24,545	84,665
1998–1999	310,738	23,275	72,530	25,706	94,846
1999–2000	315,047	22,034	89,754	26,885	94,005
2000–2001	325,448	24,588	106,523	27,860	91,264
2001–2002	343,124	24,569	110,738	28,037	93,054
2002–2003	361,745	26,029	115,033	29,963	83,404
2003–2004	379,405	28,477	123,807	30,906	93,022
2004–2005	404,691	30,087	129,842	30,775	101,621
2005–2006	433.27[2]	66.84[2]	130.11[2]	31.33[3]	592,194
2006–2007	462.35[2]	72.34[2]	146.55[2]	30.79[3]	639,008
2007–2008	502.81[2]	82.82[2]	156.10[2]	31.48[3]	689,780
2008–2009	549.57[2]	89.19[2]	160.77[2]	31.75[3]	712,540
2009–2010	585.30[2]	105.21[2]	192.77[2]	46.52[3]	761,934
2010–2011	589.87[2]	102.2[2]	196.99[2]	51.25[3]	809,455
2011–2012	638.84[2]	139.44[2]	204.12[2]	46.48[3]	811,506
2012–2013	704.34[2]	46.05[2]	219.21[2]	39.78[3]	908,574
2013–2014	722,568	44,637	227,027	52,373	974,436
2014–2015	827,516	49,584	226,896	51,300	1,054,355
2015–2016	847,576	45,475	239,792	52,517	1,104,228
2016–2017	858,576	47,318	249,941	55,697	1,163,290
2017–2018	896,086	46,977	256,117	59,170	1,232,505
2018–2019	958,250	42,724	260,701	60,798	1,307,685
2019–2020	1,001,853	42,234	259,124	64,144	1,323,048
2020–2021 (P)	955,054	35,876	228,605	60,645	1,315,110

GWh: Giga Watt hour = 106 × Kilowatt hour. Abbr.: P: Provisional.

Source: Ministry of Statistics and Programme Implementation, Govt. of India. (ON1660) & Past Issues; Ministry of Petroleum & Natural Gas, Govt. of India (ON2933 & past issues. Data Source: indiastat.com. Data accessed on 25 July 2023.

Table 1.14 Consumption of Primary Sources of Conventional Energy in India (1970–1971 to 2020–2021)

Year	Coal (in '000 tonne)	Lignite (in '000 tonne)	Crude petroleum[2] (in '000 tonne)	Natural gas (in million cubic metres)	Electricity[1] (in GWh)
1970–1971	71,230	3,390	18,379	647	43,724
1971–1972	74,060	3,730	20,042	718	47,073
1972–1973	78,180	2,890	19,328	771	49,088
1973–1974	77,660	3,320	20,958	762	50,246
1974–1975	85,580	2,940	21,094	951	52,632
1975–1976	92,160	3,030	22,283	1,126	60,246
1976–1977	100,100	4,010	22,995	1,381	66,639
1977–1978	103,780	3,590	24,898	1,464	69,255
1978–1979	100,150	3,270	25,974	1,711	77,293
1979–1980	105,530	3,050	27,474	1,681	78,084
1980–1981	109,310	5,100	25,836	1,522	82,367
1981–1982	121,010	6,310	30,146	2,222	90,245
1982–1983	130,130	6,980	33,156	2,957	95,589
1983–1984	137,290	7,245	35,263	3,401	102,344
1984–1985	141,450	7,839	35,556	4,141	114,068
1985–1986	155,530	7,675	42,910	4,950	123,099
1986–1987	166,840	8,027	45,699	7,075	135,952
1987–1988	179,209	8,103	47,754	7,968	145,613
1988–1989	192,115	8,542	48,803	9,250	160,196
1989–1990	203,424	10,159	51,942	11,172	175,419
1990–1991	213,360	14,200	51,772	12,766	190,357
1991–1992	232,330	15,881	51,423	14,442	207,645
1992–1993	241,750	15,726	53,482	16,116	220,674
1993–1994	256,320	17,889	54,296	16,340	238,569
1994–1995	269,174	20,090	56,534	17,337	259,629
1995–1996	284,037	22,296	58,741	18,091	277,029
1996–1997	298,620	22,402	62,870	18,632	280,146
1997–1998	306,824	22,980	65,166	21,513	296,749
1998–1999	313,476	23,367	68,538	22,489	309,734
1999–2000	315,047	21,804	85,964	26,885	312,841
2000–2001	339,306	24,824	103,444	27,860	316,600
2001–2002	349,584	24,578	107,274	28,037	322,459
2002–2003	361,833	26,010	112,559	29,964	339,598
2003–2004	379,283	28,486	121,841	30,906	360,937
2004–2005	407,411	30,087	127,117	30,775	386,134
2005–2006	407.04[3]	30.23[3]	130.11[4]	26.86[5]	411,887
2006–2007	462.35[3]	30.81[3]	146.55[4]	37.60[5]	455,749
2007–2008	502.82[3]	34.65[3]	156.10[4]	39.80[5]	501,977
2008–2009	549.57[3]	31.85[3]	160.77[4]	32.99[5]	553,995
2009–2010	585.30[3]	34.41[3]	186.55[4]	48.34[5]	612,645
2010–2011	589.87[3]	37.69[3]	196.99[4]	52.02[5]	694,392
2011–2012	642.64[3]	41.89[3]	204.12[4]	60.68[5]	785,194
2012–2013	688.75[3]	46.01[3]	219.21[4]	53.91[5]	824,301
2013–2014	739,342	43,897	222,497	52,375	874,209
2014–2015	822,131	46,954	223,242	51,300	948,522
2015–2016	836,727	42,211	232,865	52,517	1,001,191
2016–2017	837,220	43,155	245,362	55,697	1,061,183
2017–2018	898,519	46,317	251,935	59,170	1,123,427
2018–2019	968,253	45,811	257,205	60,798	1,209,972
2019–2020	950,106	42,317	254,386	64,144	1,248,086
2020–2021 (P)	905,883	–	221,773	60,645	1,237,800

GWh: Giga Watt hour = 106 × Kilowatt hour. Abbr.: P: Provisional. Source: Ministry of Statistics and Programme Implementation, Govt. of India (ON2703 & past issues); Ministry of Petroleum & Natural Gas, Govt. of India (ON2933 & past issues). Data Source: indiastat.com. Data accessed on 25 July 2023.

In addition to coal, India boasts substantial mineral wealth, encompassing minerals such as iron ore, bauxite, copper, zinc, lead, limestone, and many others. These minerals play a vital role in various industries, including manufacturing, construction, infrastructure development, and renewable energy.

India's mineral reserves are distributed across different regions, with notable mining areas situated in states like Odisha, Jharkhand, Rajasthan, Gujarat, and Karnataka. The mining sector has experienced significant growth, attracting both domestic and foreign investments. The government has implemented policies to foster responsible and sustainable mining practices, promoting efficient resource utilization while minimizing environmental impacts.

The coal and mineral reserves in India are instrumental in driving the nation's economic growth, generating employment opportunities, and meeting energy demands. The government remains committed to exploring new mining prospects, adopting advanced technologies, and striking a balance between resource extraction and environmental conservation. This strategic approach aims to harness the potential of India's coal and minerals sector, fostering sustainable development, and addressing the challenges associated with resource depletion and environmental degradation.

India is fifth in the world in terms of proved coal reserves, with 107,727 Mt. This accounts for roughly 9% of the world's total coal reserves, which are estimated to be 1,139,471 Mt. India's proven reserves are equivalent to 111.5 times its annual coal consumption. Based on current consumption levels and the elimination of untested reserves, India is estimated to have around 111 years of coal reserves remaining. In 2016, India's annual coal consumption was 966.3 Mt (https://www.worldometers.info/coal/india-coal/).

At the same time, India is the world's second largest consumer of coal, accounting for approximately 84.8% of total global consumption of 1,139,471,430 tonnes.

India utilizes 729,540 cubic feet of coal per capita per year (based on a 2016 population of 1,324,517,249), or 1,999 cubic feet per capita per day. Between 1990 and 2000, India saw a 48.67% increase in coal usage. It consumed around 4.55% of the coal as compared to the world's coal consumptions in 1990 (Table 1.15 and Figure 1.5).

Table 1.15 Consumption of Coal (History and Projection) in India, USA, and the World (1990, 2000, 2001, 2005, 2010, 2015, 2020, and 2025).

(in million tonne)

Region/ Country	History			Projections				
	1990	*2000*	*2001*	*2005*	*2010*	*2015*	*2020*	*2025*
India	219	325.60	326.50	346.40	390.90	428.10	462	526
USA	819	983.20	961.40	1,003.10	1,104.70	1,162.80	1,231.70	1,309.70
World Wide	4,813.40	4,639.30	4,773.50	4,971.20	5,462.80	5,862.80	6,266.40	6,786.10

Source: Consumption of Coal (History and Projection) in India, USA and World (1990, 2000, 2001, 2005, 2010, 2015, 2020, and 2025). Data Source: indiastat.com. Data accessed on 25 July 2023.

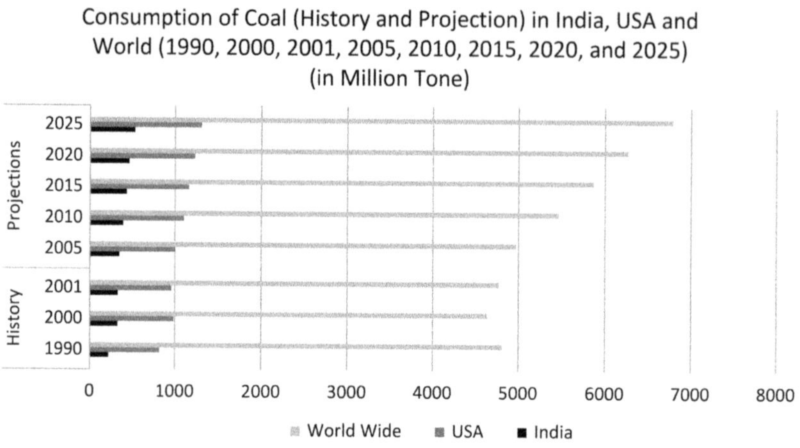

Consumption of Coal (History and Projection) in India, USA and
World (1990, 2000, 2001, 2005, 2010, 2015, 2020, and 2025)
(in Million Tone)

World Wide   USA   India

Figure 1.5 Consumption of Coal (History and Projection) in India, USA, and the
World (1990, 2000, 2001, 2005, 2010, 2015, 2020, and 2025).

It is projected that by 2025 the coal consumption of India will be 526 million tonnes, which is 7.75% as compared to the world's coal consumption. India's demand for coal consumption is increasing at an average rate of 5.5%. In
2019–2020, it was predicted that India will require 1,000 million tonnes of coal
in order to meet its requirements. It is predicted that India will need 1,273 million tonnes of coal for its daily operations (Table 1.16 and Figure 1.6).

The stock of raw coal pit has gone through a fluctuating growth. Starting from 1996–1997 to 2001–2002, it has been through a downfall;
2002–2003 to 2011–2012, it has witnessed a slow rate of growth. Immedi-
ately after 2011–2012, it has been through a downfall till 2014–2015. Now a
steady increase in the stock of raw coal pits has been noticed. Highest growth
has been identified in the year 2019–2020, i.e., 41.28% as compared to its
previous year (Table 1.17 and Figure 1.7).

Table 1.16 Projection of Coal Demand in India
(2019–2020 to 2023–2024)

Year	*All India demand (in million tonne)*
2019–2020	1,000
2020–2021	1,052
2021–2022	1,111
2022–2023	1,172
2023–2024	1,273

Note: As Per Vision 2024 of Ministry of Coal.

Source: Rajya Sabha Unstarred Question No. 2406,
dated on 16 March 2020.

Data Source: indiastat.com. Data accessed on 25
July 2023.

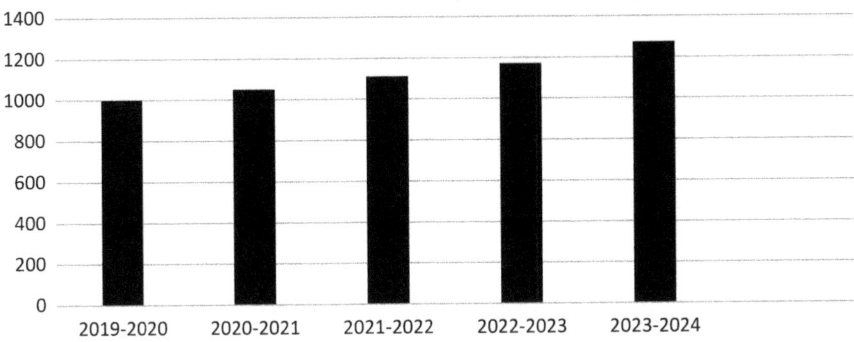

Figure 1.6 Projection of Coal Demand in India (2019–2020 to 2023–2024).

Table 1.17 Share of Raw Coal Pit-Head Closing Stock in India
(1996–1997 to 2019–2020)

(in million tonne)

Years	Quantity	Growth (%age)
1996–1997	42.02	–7.93
1997–1998	41.47	–1.33
1998–1999	40.10	–3.30
1999–2000	29.32	–26.88
2000–2001	21.43	–26.90
2001–2002	18.12	–15.43
2002–2003	19.39	7.01
2003–2004	21.28	0.74
2004–2005	23.97	12.62
2005–2006	34.33	43.24
2006–2007	44.35	29.42
2007–2008	46.78	–
2008–2009	47.32	–
2009–2010	64.86	–
2010–2011	72.19	11.30
2011–2012	74.04	2.56
2012–2013	63.05	–14.84
2013–2014	55.51	–11.95
2014–2015	59.39	6.98
2015–2016	65.36	10.06
2016–2017	76.89	17.64
2017–2018	62.04	–19.32
2018–2019	57.64	–7.09
2019–2020	81.43	41.28

Source: Ministry of Coal, Govt. of India (ON2723 & past issues).
Data Source: indiastat.com. Data accessed on 25 July 2023.

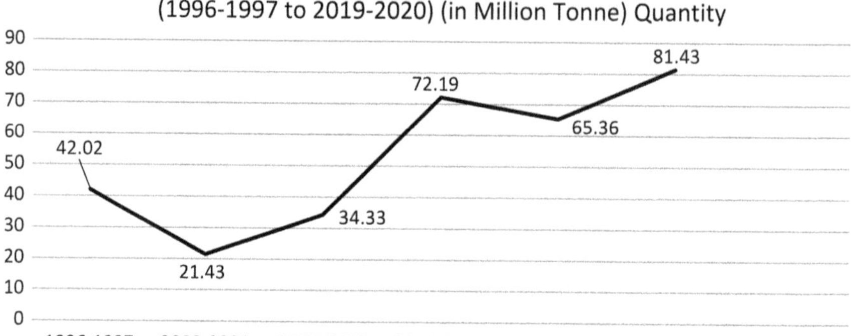

Figure 1.7 Share of Raw Coal Pit-Head Closing Stock in India (1996–1997 to 2019–2020).

Mine head stocks of India was at 72,192,000 million tonnes in during 2011–2012, in 2018–2019 it was estimated that we are left with 57,640,000 million tonnes of coal head stock. Jharkhand, Odisha, Chhattisgarh, Madhya Pradesh, and Maharashtra were having the largest head stock. At the beginning of 2011–2012 Jharkhand had the head stock reserves of 27,128,000 million tonnes of head stock, at the end of 2011–2012 it was reduced to 24,684,000 million tonnes, states like Odisha, Madhya Pradesh, and Maharashtra experienced a reverse trend in 2011–2012. At the beginning of the year, Odisha, Madhya Pradesh, and Maharashtra were left with 21,611,000 million tonnes, 4,391,000 million tonnes, and 3,793,000 million tonnes, respectively, but at the end of 2011–2012 an increase in their head stock was found. The closing stock of head stock was increased by 650,000 million tonnes, 1,874,000 million tonnes, and 1,048,000 million tonnes, respectively. At the end of 2018–2019, Jharkhand (20,645,000 million tonnes), Odisha (11,178,000 million tonnes), Maharashtra (10,917,000 million tonnes), Chhattisgarh (7,359,000), and Telangana (4,921,000 million tonnes) were the largest holder of coal head stock in India (Table 1.18).

East Coalfields Ltd (ECL) recovered the largest amount of coal extracted from illegal mining. In 2016–2017, it recovered 973.90 million tonnes of coal, i.e. worth Rs.48.69 lakhs from its West Bengal branch and 285.50 tonnes worth Rs.13.60 lakhs from its Jharkhand branch. In 2018–2019, it recovered 274.11 million tonnes of coal from the West Bengal branch and 3 tonnes of coal from its Jharkhand branch. Central Coalfield, situated in Jharkhand, recovered 33 tonnes of coal in 2016–2017. The total illegal coal recovered in 2016–2017 was 1292.40 tonnes of Rs.62.62 lakhs. In 2018–2019, it has reduced to 277.11 tonnes of Rs.13.85 lakhs. In 1996–1997, pit-head closing stock of India was at 42.02 million tonnes, which shared 99 per cent shares in total solid fossil fuels by –7.93% change over previous year. However, this closing stock started decreasing up to 2003–2004. In 2003–2004, it was 21.29 million tonnes covering 99.01%

Table 1.18 Pit-Head Closing Stock of Different Solid Fossil Fuels in India (1996–1997 to 2020–2021)

(in million tonne)

Year	Raw coal			Lignite			Total solid fossil fuel	
	Pit-head closing stock	Share in total solid fossil fuel (%age)	Change over previous year (%age)	Pit-head closing stock	Share in total solid fossil fuel (%age)	Change over previous year (%age)	Pit-head closing stock	Change over previous year (%age)
1996–1997	42.02	99.00	–7.93	0.43	1.00	69.05	42.45	–7.51
1997–1998	41.47	99.39	–1.33	0.25	0.61	–40.38	41.72	–1.72
1998–1999	40.10	99.02	–3.30	0.40	0.98	56.69	40.49	–2.94
1999–2000	29.32	97.22	–26.88	0.84	2.78	110.80	30.16	–25.53
2000–2001	21.43	97.73	–26.90	0.50	2.27	–40.64	21.93	–27.28
2001–2002	18.12	96.07	–15.43	0.74	3.93	49.00	18.87	–13.97
2002–2003	19.39	96.37	7.01	0.73	3.63	–1.48	20.13	6.67
2003–2004	21.29	99.01	9.78	0.21	0.99	–71.00	21.50	6.85
2004–2005	23.97	97.81	12.58	0.54	2.19	152.83	24.50	13.96
2005–2006	34.33	98.49	43.24	0.53	1.51	–2.05	34.86	42.25
2006–2007	44.35	97.79	29.17	1.00	2.21	90.86	45.35	30.10
2007–2008	46.78	99.30	5.48	0.33	0.70	–67.27	47.11	3.87
2008–2009	37.32	98.13	1.15	0.90	1.87	175.30	48.22	2.36
2009–2010	64.86	99.14	37.08	0.56	0.86	–37.43	65.43	35.69
2010–2011	72.19	99.16	11.30	0.61	0.84	7.96	72.80	11.27
2011–2012	74.04	98.60	2.56	1.05	1.40	72.30	75.09	3.14
2012–2013	63.05	97.69	–14.84	1.49	2.31	42.06	64.54	–14.05
2013–2014	55.51	96.76	–11.95	1.86	3.24	24.58	57.37	–11.11
2014–2015	59.39	94.92	6.98	3.18	5.08	70.75	62.57	9.05
2015–2016	65.36	93.15	10.06	4.81	6.85	51.42	70.17	12.16
2016–2017	76.89	91.78	17.64	6.88	8.22	43.13	83.77	19.38
2017–2018	62.04	89.59	–19.32	7.21	10.41	4.75	69.25	–17.34
2018–2019	57.64	91.04	–7.09	5.67	8.96	–21.33	63.31	–8.57
2019–2020	81.43	93.68	41.28	5.50	6.32	–3.12	86.93	37.30
2020–2021	109.06	95.63	33.93	4.98	4.37	–9.35	114.04	31.19

Note: Large variation in closing stock is due to adoption of book stock as opening stock and vendible stock as closing stock in case of CCL during the year 1999–2000 while the same during 2000–2001 is for write-off of stock (CIL).

Source: Ministry of Coal, Govt. of India (ON2875 & past issues).

Data Source: indiastat.com. Data accessed on 25 July 2023.

of total solid fossil fuels by 7.01% over its previous year. Then it started increasing till 2020–2021 and currently we are left with 109.06 million tonnes which covers 95.63% of total fossil fuels by 33.93% change over its previous year.

1.7   Natural Capital: Issues of Sustainability

The current situation of natural capital in India reveals a complex and diverse landscape. The agricultural sector plays a crucial role, with varying landholdings and net sown areas across different regions. The production of food, cereals, and crops showcases the country's agricultural prowess. Food sufficiency in the country is largely achieved through bumper production of cereals such as rice and wheat. There is a consistent surplus in production of these two cereal crops over the last decade which enabled the government of India to manage a massive scheme of distributing free rice and wheat to about 800 million people in the country since the pandemic broke out. However, as the net sown area in the country is stagnant and even showing a slow declining trend, the productivity of crops takes the centerstage in food security debate. Apart from cereal crops, the productivity in vegetables, fruits, and pulses is dwindling and is the major reason for food inflation in the country recently. Climate change poses major threats for agricultural production as extreme weather events such as heatwaves, erratic rainfall, and floods and droughts have damaged crops in a regular basis. Such turn of events calls for a major rethinking in crop production in the country. There is a need for fostering sustainable agricultural practices to conserve the natural capital in India. The practices such as crop diversification, agroforestry, maintaining organic cover in the crop fields, minimum tillage and precision in the use of chemical fertilizer and pesticide have been identified as sustainable farming practices that can help conserve biodiversity.

Forests and fisheries contribute significantly to the nation's ecological wealth, while the carbon stock and sinking capacity of Indian forests underline their importance in mitigating climate change. According to the official statistics of the government of India, there is an increase in forest cover in the country (ISFR, 2021). However, several experts have flagged concerns about this trend. The definition of the forest has evolved to include the commercial plantations as part of the forest. Studies show that the native forests, which are the core of tropical rainforests, have been declining in India (Puyravaud et al., 2010; Roy and Fleischman, 2022). The official statistics do not delineate native forests from tree plantations, although both provide different ecosystem services. The native forests are rich with biodiversity and provide more sustainable outcomes such as carbon sink services. The evidence shows that native forests in the Western Ghats declined at a mean rate of 0.99% per year from 1973 to 1995 (Jha et al., 2000), whereas those of the Himalayas shrank at 1.18% per year between 1990 and 2000 (Pandit et al., 2007). The deforestation of native forests in central India and Tamil Nadu is even faster. So, while placing natural capital as an important policy focus, it should be considered that commercial tree plantations cannot restore the ecosystem services of the native forests.

The renewable energy sector is witnessing rapid growth, with diverse sources and increasing capacity, highlighting India's commitment to sustainable development. However, the reliance on coal and mining activities raises concerns regarding environmental impacts. Overall, this chapter provides an overview of the current status of India's natural capital, emphasizing the need for sustainable practices and policies to safeguard the country's rich biodiversity, mitigate climate change, and ensure a balanced approach towards sustainable economic growth and ecological conservation.

References

ISFR (2021) Forest Survey of India, Ministry of Environment, Forest and Climate Change. Welcome To Forest Survey of India (fsi.nic.in), Accessed on 24.06.2024

Jha, C. S., Dutt, C. B. S. and Bawa, K. S. (2000) 'Deforestation and land use changes in Western ghats, India', *Current Science*, 79(2), pp. 231–238.

Pandit, M. K., Sodhi, N. S., Koh, L. P., Bhaskar, A. and Brook, B. W. (2007) 'Unreported yet massive deforestation driving loss of endemic biodiversity in Indian Himalaya', *Biodiversity and Conservation*, 16(1), pp. 153–163. https://doi.org/10.1007/s10531-006-9038-5

Puyravaud, J. P., Davidar, P. and Laurance, W. F. (2010) 'Cryptic destruction of India's native forests', *Conservation Letters*, 3(6), pp. 390–394. https://doi.org/10.1111/j.1755-263X.2010.00141.x

Roy, A. and Fleischman, F. (2022) 'The evolution of forest restoration in India: The journey from precolonial to India's 75th year of independence', *Land Degradation and Development*, 33(10), pp. 1527–1540. https://doi.org/10.1002/ldr.4258

2 Measurement of Economic Damages of Climate Change Without Suitable Adaptation Strategies

2.1 Introduction

Anthropogenic greenhouse gas emissions have caused variations in climate patterns and the occurrence of adverse climate events (floods *and* droughts).[1] This shift in climate patterns has a direct impact on biodiversity. Past studies in the recent literature, estimating the economic impact of climate change, have turned the spotlight on evaluating climate change's impact on several indicators (agriculture yields, gross domestic product (GDP), financial markets, farmer's behaviour, migration, water, mortality) (Bandara and Cai, 2014; Carleton, 2017; Kalli and Jena, 2022a; Jena and Grote, 2022). Agriculture is the primary sector among most of the developing economies which acts as the major source of food security. With the limited availability of resources to adapt, developing nations are highly susceptible to climate change. The situation is grimmer as most of the developing economies are in tropical regions, where future projections of climate change indicate devastating distress in the agricultural sector. The climate in India can be classified as tropical wet and dry climate. The annual mean temperature showed a significant warming trend of 0.51°C for the period 1901–2007 (Revadekar et al., 2012). The long-time trend indicated decreased dispersion of southwestern and northeast rainfall in peninsular India (Varadan et al., 2015). These significant climate trends with uneven patterns of rainfall and compounded with rising temperatures will harm crop yields.

2.2 Implications of Climate Change Vulnerability to Agriculture

Climate change has a significant effect on agriculture. A small change in the climate indicators like temperature, rainfall, solar radiation and air pressure affects the plant growth, and causes a significant decline in the final yield. Of course, there are other inputs like irrigation, fertilizer, pesticides, high-yielding variety seeds, and farm management practices that influence plant growth and yield. However, unlike climate, other input parameters can be controlled by farmers, while climate acts as the primary source of input for plant growth.

DOI: 10.4324/9781003290445-2

Among different climate parameters, rainfall and temperature act as the primary variables that influence agriculture production. Rainfall is the main source of water to plant growth in rainfed agricultural systems. Variation in the total discharge in rainfall over a period of time, dry spell and wet spell affect the water source which in turn causes significant impact to cultivation. Significant delay in the onset of rainfall leads to delay in cultivation, which further affects plant physiological process. Scanty rainfall tends to decline the total output of crop, whereas drought and flood situations lead to entire crop loss; and further frequent flooding may cause salinity intrusion in crop fields. The extreme climate shocks like a rise in day's maximum temperature, erratic or variation in rain, and frequent drought pose a threat to farming systems. Temperature response has a significant impact on plant growth. Temperature helps in the physiological process such as photosynthesis in plants (Yoshida, 1981). Rise in temperature beyond the physiological limits of a plant can be harmful to it, which eventually leads to higher desiccation[2] rates.

The primary dependency of agriculture on climate without suitable adaptation processes is a significant challenge in climate change context, although development in agriculture technology offsets the climate change impact to certain extent. Absence of adequate policy formulation for adaptation leads to adverse effect of climate change on traditional farming practices. Climate change vulnerability may cause a significant effect on livelihood and food security issue in a new dimension. Agriculture is the primary sector among majority of countries acting as major source of food security for these countries' inhabitants. The negative effect of climate change on agricultural production may cause food crisis, and further lead to sharp increase in food prices. Farmers among developing countries have already experienced food insecurity and traditional livelihood problems due to climate variability (IPCC, 2007). As the world population had already reached seven billion in 2011, and the projected population for the year 2050 is nine billion, stress on agriculture sector to produce more is mounting. Even with significant technological advancement, the production of food crops is not sufficient to provide adequate nutrition to the seven billion inhabitants of the world. As a result, almost two billion people are suffering from hidden hunger all over the world. Developing economies in Sub-Saharan Africa and South Asia are the regions tormented by relatively high level of poverty and food insecurity (Alexandratos and Bruinsma, 2003). The deficiency of micronutrients is causing nearly 1.1 million child deaths every year. Most of the countries have committed themselves to reducing malnourishment and undernourishment by the end of the year 2050 (Von Grebmer et al., 2014). Apart from generating food, agriculture has served as the major source of employment in the developing economies. This raises the challenges ahead for the agriculture sector in the context of climate change to produce enough food, livelihood issues among rural areas and profitability in the agriculture sector. Thus, with climate variations increasing, suitable adaptation methods would have to be framed for the farming communities as resilience options (Adger et al., 2003).

More adverse effects can be found in the regions under tropical climate when compared to the regions with temperate climate. Tropical dry and wet climate has a prolonged dry season which often lasts several months, and the impact may be severe and may last for several months due to drought condition. The summer droughts in tropical regions (continents located in the mid-latitudes) would face wildfire, which would further lead to deforestation and destruction in the flora and fauna (Food and Agriculture Organization (FAO), 2002). Adaptation strategies of farmers to climate change in land use pattern have led to the migration of butterflies to the northern part of the United States.

2.3   Climate Change in Developing Countries: Indian Context

Climate change is expected to highly affect the developing countries due to lack of adaptation strategies. India is the second most populous country globally, with a population of 1.3 billion, and has the seventh largest land area in the world at 3.288 million square kilometres. It has a diverse ecosystem due to its range of climates, from humid and dry tropical in the south to temperate alpine in the north. Agriculture contributes 23% to India's GDP (FAO, 2010). While the green revolution has led to technological advances in the agricultural sector, food security remains a concern. Studies highlight temperature as a significant factor in climate change. Recent studies show that climate trends have adversely affected the Indian continent (Burney and Ramanathan, 2014).

The projections of Jayaraman and Murari (2014), using the data from the Indian Metrological Department, show the average temperature of India rose from 0.6°C to 0.8°C from 1850 to 2010. Figure 2.1 shows the variation in the average annual temperature from the 30-year standard during the period from

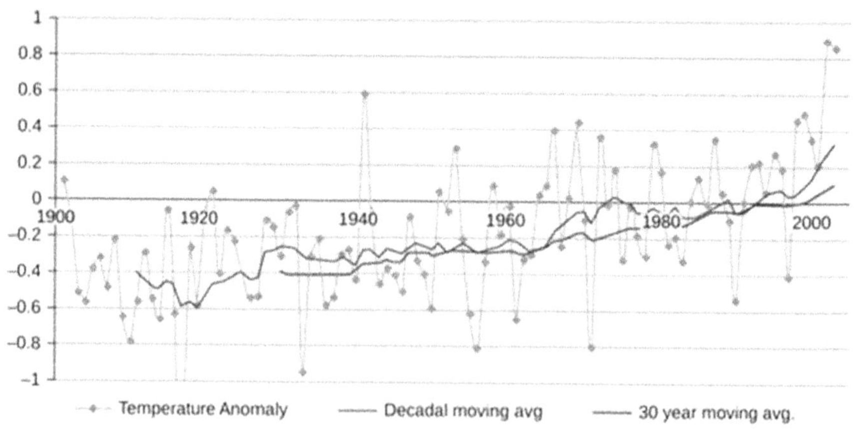

Figure 2.1 Change in Mean Temperatures, India, 1900–2009.

1960 to 1999. The trend shows an increase in the mean temperature over the period. In India, the projected increase in temperature for the future is greater in winter, followed by summer, monsoon, and post-monsoon seasons. According to the Representative Concentration Pathway (RCP) 8.5 emission scenario, rare cold and heat wave events are likely to increase by the end of the twenty-first century in India (Basha et al., 2017).

Agriculture is one of the primary sectors which plays a crucial role in India's overall economic and social well-being. But the decrease in GDP and employment over the year shows the agriculture and allied sectors are also facing the most vulnerability due to climate change (Mall et al., 2006). Food security is directly or indirectly impacted by climate change. Due to the population explosion, food scarcity is growing, which may create severe food insecurity in the nation (Alam et al., 2018). Due to climate change, food security has been considered the most prominent challenge in India. Climate change has affected the irrigated rice yields by about 10% in most coastal districts (Abeysingha et al., 2016). Abeysingha et al. (2016) analysed the rainfall and temperatures change during rice and wheat growing periods and found that climate change had a negative impact on crop production. A sectoral and regional analysis prepared by the Ministry of Forest and Environment indicates that the daily extremes in surface air temperature may intensify in the 2030s. The spatial pattern of the change in the lowest daily minimum and highest maximum temperature will alarm warming of 1–4°C towards the 2030s.

Anwar et al. (2013) found that extreme weather events and disasters adversely affect the agricultural sector of developing countries. The impact of climate change on agriculture is unpredictable and ultimately depends on various climatic factors and extreme occurrences, including droughts and floods. The impact of climate change has been categorized into two aspects: the biophysical aspect and the socioeconomic aspect (Anwar et al., 2013). As this study does not primarily focus on the impact of climate change on agriculture, the literature review has been done on the general impact of climate change on agriculture around the globe and in India. The impact of climate change literature shows two broad types of impact assessment models: "general equilibrium models" and "partial equilibrium models". However, this literature survey will be reported chronologically on the historical climate change impact studies.

Adams et al. (1990) conducted a study to investigate the potential impact of climate change and increased CO_2 on agriculture in the US. Using the predictions of global climate models (GCMs), they found that changes in temperature and precipitation could lead to decreased yields of wheat, maize, and soybean and increased water requirements for crops. Nordhaus (1993) estimated the economic effect of doubling atmospheric carbon dioxide and climate change on world agriculture. Changes in domestic yields affect the prices of agricultural commodities, and changes in global consumption and production affect economic well-being. Kaiser et al. (1993) analysed the potential economic and agronomic effects of climate change. Moderate warming had a positive impact on crop yield; extreme warming had a negative impact on yield. The increase in global mean temperature and impacts were not consistent across sectors. Some sectors exhibited increasing adverse impacts with increasing global mean temperature (GMT), particularly coastal resources, biodiversity, and possibly marine ecosystem productivity (Hitz and Smith, 2004).

Rosenzweig and Hillel (1995) examined the effects of climate change on global food production and agriculture. Physical impacts of climate change could negatively impact crop yields, pest and disease pressure, and soil fertility (Antle, 1995). The physical impact of climate change could lead to the hindrance of economic development of the region. Climate change has adverse economic and social impacts, including food security concerns and decreased agricultural productivity (Mendelsohn and Tiwari, 2000).

Rosenzweig et al. (2001) found that changes in temperature, precipitation, and extreme weather events are likely to have significant implications for food production and the prevalence and severity of plant diseases and pest outbreaks in developing countries. Mendelsohn (2008), in an insightful review of the potential effects of climate change on agricultural production in developing countries, concluded that tropical and subtropical agriculture in developing countries is more climate-sensitive than temperate agriculture. Kumar and Parikh (2001) examined the socioeconomic impact of climate change on Indian agriculture. Climate change-induced yield shocks on India's GDP could result in a decline of 1.8–3.4%, with the reduction in agricultural GDP being the major contributing factor. Kumar and Parikh (2001) examined the sensitivity of Indian agriculture to these physical impacts. A 2°C increase in temperature and a 7% increase in precipitation could result in an estimated loss of about 8.4% of the total net revenue. The study by Guiteras (2009) assessed the impact of stochastic inter-annual weather fluctuations on agricultural productivity over 40 years. The study revealed that the anticipated climate change during the timeframe of 2010–2039 is expected to result in a decrease of 4.5–9% in the yields of significant crops. Nelson et al. (2009) used a global economic model, a crop simulation model, and hydrological models to assess climate change. They projected that, by 2050, the yields of major crops in developing countries could decline by as much as 10%, leading to significant food price increases and reductions in food security. (Lobell et al., 2012) assessed the impact of extreme heat on wheat production by using nine years of satellite measurements of wheat grown in northern India and found that temperatures during the grain-filling stage of wheat growth exceeded a certain threshold and that the wheat plants died prematurely, resulting in lower yields. Gupta et al. (2017) examined the impact of climate change on wheat production in India. Rising temperatures and air pollution have contributed to a decline in wheat yields in India. Higher temperatures during the wheat growing season have reduced yields, with the negative impact being more pronounced in regions with high levels of air pollution. Abeysingha et al. (2016) examined the impact of climate change on rice and wheat production in the Gomti river basin of India. The study found that the temperatures and changes in precipitation patterns could lead to lower crop yields, with the negative impact being more pronounced for rice than wheat. Mishra et al. (2016) investigated the impact of climate change on agricultural production in Odisha, a state in eastern India. The Ricardian analysis shows higher temperatures during the growing season and increased rainfall variability are associated with lower farm-level net revenue of paddy. Bangladesh, India, and Pakistan are experiencing more frequent and severe flooding between 0°C and 2°C global mean temperature changes (Mirza et al., 2011). Bandara and Cai (2014) examined the potential impacts of climate change on food crop productivity, food prices, and food security in South Asia, namely Bangladesh, India, Nepal, Pakistan, and Sri Lanka. Tim Wheeler (2013) reviewed the aspects of climate change and food security. The effect of climate change on crop productivity could have a negative impact on progress towards ending world hunger, with potential consequences for food availability.

2.4 Impact of Climate Change on Agricultural Sector: A Case Study

The present study has attempted to estimate the impact of climate change on agriculture's net revenue and crop yield. This information can be used for policy formulation to reduce the loss, and to frame suitable adaptation strategies to boost agricultural production. The classification of the region's climate vulnerability with the inclusion of all the possible factors will provide the information to frame coping mechanisms to overcome the climate change effect. The study employed a fixed effect panel regression model to estimate the climate change impact on net revenue per hectare and crop yields. Panel data, with a cross-section of 20 districts and a period of 21 years (1992–2012) was considered in the study. The strategy is to link the historical climate data that has significant variation to the revenue and crop yield, to examine the overall and individual effects in the agriculture sector. Finally, the result from the study has been used to draw the policy implications.

Karnataka belongs to the tropical zone in India which follows a semi-arid climate. The state lies in the southwestern part of India with an area of 1.92 lakh km² and a population of sixty-one million. In the present study, we fix the growing season to months from June to September as Kharif is the predominantly significant cultivation period. Several episodes of high-intensity drought and significant variations in monsoon rainfall have been observed in Karnataka (Guhathakurta et al., 2015). Agriculture in Karnataka is highly dependent on the monsoon rainfall and highly susceptible to climate change (Kalli and Jena, 2022b); 26.5% area is irrigated while the remaining depends on Rainfed cultivation (Bhende and Tembhare, 2013). The state has a varied topographical character ranging from coastal plains to gentle slopes and the heights of the Western Ghats. There is a need to perform a consistent assessment of climate change's impact on crop yields. This analysis will exhibit the impact of climate change in arid and semi-arid regions. The dearth of studies in assessing climate change impact at a regional scale can be noted. In this context, the present assessed the region-specific climate sensitivities on crop yields using statistical techniques. The analysis from the present study contributes to the literature about climate change impact assessment on agriculture and also to formulate policy measures in the region.

2.5 Methodological Issues

The present study provides a detailed empirical analysis of climate change's impact on agriculture in Karnataka, a south Indian state. The state is a semi-arid region with significant variation in the climate characteristics. Several episodes of high-intensity drought and significant decreasing trends of monsoon rain have been observed in Karnataka (Guhathakurta et al., 2015). The objectives in the present study are three-pronged. First, the study assesses the impact of climate change on the agriculture net revenue per hectare. The agrarian rural economy in the Indian states does not possess diversified income, where the primary source of income is dependent on agriculture. To consolidate the incomes of farmers from agriculture in the face of climate change, a thorough understanding of the dynamic relationship between agriculture and climate change is required. Keeping that in view, the first objective has attempted to measure the impact of climate change on net agricultural revenue. Secondly, the study assesses the climate change impact on the major crop yields in the kharif season. The rising food insecurity and growing population are the two major issues that can be addressed successfully with an increase in agriculture productivity, which requires in-depth knowledge of the relationship between crop yields and changes in climate trends. Third, the study maps the districts in the state using a metric of climate vulnerability considering the agriculture sector as the prime focus. This vulnerability assessment helps in identifying the regions that are most vulnerable having more exposure to climate change and less adaptive capacity that need immediate policy attention. Climate change has a significant impact on the human-managed system, while each sector will be affected differently, where the outcome of the result and interpretation vary with the scope of the vulnerability assessment.

2.6 Impact of Climate Change on Net Revenue

The present study has adopted a fixed effect panel regression model to establish the empirical relationship between the net revenue per hectare and the climate variables. The study includes 20 districts of Karnataka as the cross-sectional units and the period used spans from 1992 to 2012. A fine-scale gridded dataset of $0.25° \times 0.25°$ for rainfall and $1° \times 1°$ for temperature was used to construct the district-wise climate data based on their boundaries. Agriculture data was obtained from the Directorate of Economics and Statistics (DES), Karnataka. Linear trend and irrigation are used as the additional independent variables in the estimation.

The net revenue per hectare is the function of climate variables, where season-wise climate was classified and used as the independent variable. Kharif season (June to September), Rabi season (October to December), post-Rabi (January and February), and summer season (March to May) are the four different seasonal climates. Three models were estimated and compared in terms of their impact to understand the climate change impact on the net revenue (see Table 2.1). Model 1 includes maximum temperature and rainfall which indicated that temperature in the kharif season had a significant negative effect on net revenue and kharif rainfall had a significant positive impact. However, other seasonal climate variables in Model 1 were statistically insignificant. In Model 2, the seasonal minimum temperature variables

Table 2.1 Fixed Effect Regression Result of Net Revenue

Variable	Model 1	Model 2	Model 3
Kharif rain	0.00099**	0.00055	0.0055***
	(0.0004)	(0.0003)	(0.001)
Rabi rain	0.00079	−0.00079	0.0015
	(0.0015)	(0.0013)	(0.003)
Kharif rain-squared	–	–	−1.06E−07***
	–	–	(2.63E−08)
Rabi rain-squared	–	–	−2.31E−07
	–	–	(0.0000003)
Maximum temperature – kharif	−0.170***	−0.218***	−0.197***
	(0.039)	(0.058)	(0.056)
Maximum temperature – rabi	−0.0566	−0.083	−0.055
	(0.0733)	(0.075)	(0.079)
Maximum temperature – post rabi	−0.0526	0.0098	−0.0051
	(0.041)	(0.041)	(0.04)
Maximum temperature – summer	0.011	−0.030	−0.048
	(0.029)	(0.035)	(0.034)
Minimum temperature – kharif	–	0.199**	0.154*
	–	(0.079)	(0.08)
Minimum temperature – rabi	–	−0.0040	0.01
	–	(0.028)	(0.02)
Minimum temperature – post rabi	–	−0.0598***	−0.058***
	–	(0.019)	(0.019)
Minimum temperature – summer	–	−0.0251	0.0082
	–	(0.0429)	(0.046)
Linear trend	0.080***	0.081***	0.078***
	(0.005)	(0.005)	(0.006)
Irrigation	0.0206*	0.018*	0.018*
	(0.01)	(0.01)	(0.01)
Constant	15.30***	14.57***	13.87***
	(3.31)	(3.24)	(3.33)
No. of observation	420	420	420
R^2	0.72	0.75	0.76

Note: Dependent variable (Net Revenue) is expressed in natural logarithm, and all exogenous variables are expressed in their physical form. Values in parentheses include robust standard errors.
*Significant at 10%; **Significant at 5%; ***Significant at 1%.

were added to the already existing variables of Model 1. Maximum temperature and rainfall showed similar results as has been observed in Model 1. However, the minimum temperature in the kharif season had a positive impact on net revenue, while in the post-Rabi season, minimum temperature had a negative impact on the net revenue. The other two seasonal minimum temperature variables were statistically insignificant. In Model 3, quadratic terms of rainfall have been added to the previous model. The results of temperature variables are similar to the previous estimation. The rainfall in the kharif season showed an inverted-U shape relationship with net revenues per hectare, whereas the quadratic term of kharif rainfall showed a negative impact on the net revenue. This can be interpreted as an increase in kharif rainfall increases profits up to a certain threshold, above which the increase in precipitation leads to losses.

The other exogenous variables namely trend and irrigation had a significant positive impact on the net revenue. Further, the study estimated out-of-sample forecasts to understand the prediction behaviour of each model. Model 2 resulted in the lowest RMSE value when compared to the other two models indicating that the former in this context provides the most accurate predictions of net revenue conditional on climate variables. In general, the results from all three models corroborate the fact that rainfall has a significant positive impact on the net revenue while rise in temperature had a significant negative impact. Adding together the marginal seasonal coefficients of maximum temperature in Model 2, the annual net revenue showed a loss of 31.3% in Karnataka. This extent of loss from the current study is much higher than the previous estimates in Indian scenario revealing a strong association of increasing temperature and loss in net revenue.

2.7 Impact of Climate Change on Crop Yields

In this section, we describe the impact of climate change on the major crop yields. Three crops pertaining to kharif season were considered for the study, namely rice, maize and finger millet. The districts that substantially cultivate these crops as indicated by the proportion of area under each crop have been considered for the study (see Table 2.2). The study estimates

Table 2.2 Districts Used in the Estimation for Rice, Maize, and Finger Millet

Type of crop	*Districts considered for the study*
Rice	Gulbarga, Raichur, Belgaum, Dharwad, Bellary, Shivamogga, U.Kannada, Chikkamagaluru, Hassan, Tumkur, D.Kannada, Mysore, Mandya, Kodagu
Maize	Raichur, Bijapura, Belgaum, Dharwad, Bellary, Shivamogga, Chitradurga, Hassan, Kolar, Mysore
Finger Millet	Shivamogga, Chitradurga, Chikkamagaluru, Hassan, Tumkur, Mysore, Mandya, Bangalore Urban, Bangalore Rural, Kolar

Table 2.3 Descriptive Statistics for Rice, Maize, and Finger Millet Datasets

Variable	Mean	Std. Dev.	Min	Max
Rice crop				
Yield (kg/hectare)	2,524.95	737.72	369.00	4,249.00
Maximum temperature (°C)	28.73	2.00	25.73	33.35
Minimum temperature (°C)	20.81	1.51	17.49	23.16
GDD (thousands)	25.19	2.27	21.62	30.06
Rain (cm)	128.55	115.42	11.42	466.70
Irrigated land (%)	59.45	40.53	0.00	100.00
Fertilizer (kgs/hectare)	123.09	62.32	15.12	334.31
Maize crop				
Yield (kg/hectare)	3,039.28	664.13	1,124.00	4,945.00
Maximum temperature (°C)	29.34	1.65	25.73	33.12
Minimum temperature (°C)	21.02	1.37	17.50	23.08
GDD (thousands)	259.49	18.9	216.20	300.60
Rain (cm)	65.3	51.07	11.27	277.62
Irrigated land (%)	43.04	35.50	0.00	100.00
Fertilizer (kgs/hectare)	115.42	49.85	27.66	284.81
Finger millet crop				
Yield (kg/hectare)	1,618.29	446.07	616.00	3,069.00
Maximum temperature (°C)	28.37	1.54	25.73	31.35
Minimum temperature (°C)	19.97	1.10	17.50	21.90
GDD (thousands)	248.77	18.58	216.20	283.80
Rain (cm)	73.92	61.23	15.01	321.44
Fertilizer (kgs/hectare)	123.12	63.18	24.11	388.74

climate yield response function using two different definitions of temperature, i.e. average maximum and minimum temperature in first definition and growing degree days in the second definition (see Table 2.3). Fixed effect panel regression approach was adopted to estimate the climate change on crop yields. Further, the study also assesses the difference in climate change impact on crop yields between irrigated and rainfed and cultivation. As a next exercise, the entire cropping season of all three crops has been split into three different growing phases, and the climate variables in each growing phase have been regressed on the crop yields to ascertain the extent of impact on these different growing phases. Results from these exercises are expected to generate in-depth information for policy planning. For all three crops two specifications of temperature have been used. The first specification includes maximum and minimum temperature and the second includes growing degree days. Model 1 includes only climate variables and Model 2 includes climate variables and other input variables, while they use maximum and minimum temperature as indicators of temperature. Model 3 includes only climate variables and Model 4 includes both climate variables

Table 2.4 Fixed Effect Regression Result of Rice Crop

Variable	Model 1	Model 2	Model 3	Model 4
Rainfall	0.0011**	0.0010**	0.0011**	0.0009**
	(0.0004)	(0.0004)	(0.0005)	(0.0004)
Max. temperature	−0.074*	−0.044*	−	−
	(0.039)	(0.025)	−	−
Min. temperature	0.003	−0.004	−	−
	(0.041)	(0.029)	−	−
GDD	−	−	−0.065*	−0.051*
	−	−	(0.036)	(0.027)
Irrigation	−	0.007**	−	0.007**
	−	(0.003)	−	(0.003)
Fertilizer	−	0.0009*	−	0.0009*
	−	(0.0005)	−	(0.0005)
Linear trend	−	0.002	−	0.002
	−	(0.004)	−	(0.004)
Constant	9.683***	8.410***	9.258***	8.355***
	(1.228)	(0.975)	(0.885)	(0.588)
No. of observation	294	294	294	294
R²	0.72	0.76	0.72	0.76

Note: Dependent variable (Yield) is expressed in natural logarithm, and all exogenous variables are expressed in their physical form. Values in parentheses include robust standard errors. *Significant at 10%; **Significant at 5%; ***Significant at 1%.

and other input variables, while temperature is represented by growing degree days in these models (see Table 2.4).

The estimates for all three crops produced significant negative effect of temperature on crop yields. The loss of crop yield indicated in the estimation with only climate variables (without addition of input variables) was significantly high among all three crops. In case of rice, maximum temperature had a significant negative effect, whereas rainfall had a positive effect. Minimum temperature effect is found to be modest and was statistically insignificant. Addition of other input variables has reduced the effect of temperature on rice yield to 4.4% from 7.4%. Similar results were obtained with the growing degree days' specification. In case of maize, maximum temperature had a significant negative effect on yield, minimum temperature was positive and statistically insignificant (see Table 2.5). Rainfall had a significant positive effect on maize yields. The result from growing degree day specification showed that each additional degree day above 34°C declines the maize yield by 10.5%, however the addition of control variables reduced the temperature effect to 5%.

Among the three crops finger millet has experienced higher crop loss due to steady rise in temperature when compared to rice and maize. The coefficient of maximum temperature resulted in a negative value of −0.26, indicating a decline in millet yields by 26% with 1°C increase in day temperature (see Table 2.6). The effect of maximum temperature declined to 22% from 26%, with the addition of the other input variables. The negative impact was found to be 18%

Table 2.5 Fixed Effect Regression Result of Maize Crop

Variable	Model 1	Model 2	Model 3	Model 4
Max. temperature	−0.097*	−0.053		
	(0.051)	(0.045)		
Min. temperature	0.011	0.005		
	(0.060)	(0.051)		
GDD	−	−	−0.10***	-0.05**
			(0.031)	(0.026)
Rainfall	0.0009*	0.0013***	0.0007*	0.0013***
	(0.0004)	(0.0004)	(0.0004)	(0.0003)
Irrigation		0.005***		0.005***
		(0.0013)		(0.001)
Fertilizer		0.002***		0.002***
		(0.0006)		(0.0006)
Linear trend		−0.01***		−0.01***
		(0.003)		(0.003)
Constant	10.5	9.01***	10.67***	9.1***
	(0.87)	(0.69)	(0.84)	(0.73)
No. of observation	210	210	210	210
R^2	0.36	0.21	0.37	0.22

Note: Dependent variable (Yield) is expressed in natural logarithm, and all exogenous variables are expressed in their physical form. Values in parentheses include robust standard errors. *Significant at 10%; **Significant at 5%; ***Significant at 1%.

Table 2.6 Fixed Effect Regression Result of Finger Millet Crop

Variable	Model 1	Model 2	Model 3	Model 4
Max. temperature	−0.26***	−0.22***		
	(0.042)	(0.043)		
Min. temperature	0.15***	0.14***		
	(0.045)	(0.043)		
GDD			−0.18***	−0.15***
			(0.029)	(0.031)
Rainfall	0.0008	0.0007	0.0008	0.0006
	(0.0008)	(0.0008)	(0.0008)	(0.0008)
Fertilizer		0.0016***		0.0018***
		(0.0003)		(0.0003)
Linear trend		−0.006*		-0.006
		(0.0025)		(0.003)
Constant	11.60***	10.71***	11.87***	10.98***
	(0.79)	(0.90)	(0.78)	(0.80)
No. of observation	210	210	210	210
R^2	0.45	0.48	0.43	0.46

Note: Dependent variable (Yield) is expressed in natural logarithm, and all exogenous variables are expressed in their physical form. Values in parentheses include robust standard errors. *Significant at 10%; **Significant at 5%; ***Significant at 1%.

without the control variables and 15% with the control variables when growing degree days was used as temperature indicator.

Growing degree days as a climate indicator has been used only in few of the past studies in the Indian context (Gupta et al., 2017). The current study has used both the average of maximum and minimum temperature as well as the growing degree days in two different specifications to verify whether the results differ. The result was similar among both the specifications for each crop. The predictive power of all four models was estimated for each crop. Model 1 with only climate variables (maximum and minimum temperature) is found to be the best-fit model in case of maize and finger millet yields. In case of rice, the model with growing degree days and the input variables was the better-fit model.

The study further investigates disaggregated assessment of climate change impact on irrigated and rainfed cultivation. Two models were estimated for both rice and maize, one with irrigated cultivation and the other with rainfed and cultivation (see Table 2.7). The results were consistent with the prior expectation that irrigation will reduce the risk of heat stress on the yield. A significant difference can be noted in two models for both rice and maize yields. The result under irrigated cultivation showed negative coefficient of –0.008 for rice and –0.047 for maize yields. However, the coefficient was statistically insignificant for rice and statistically significant for maize. The negative effect of temperature under dry land cultivation is highly devastating when compared to cultivation of rice under irrigation. The result under rainfed cultivation showed negative coefficient of –0.18 for rice and –0.13 for maize yields. The loss in the final yield is associated with the heat and moisture stress over the growing stages of plant growth. In order to assess this impact, we split the growth stage into three phases and estimate the impact of climate change on crop phenology. Vegetative phase, reproductive phase, and grain-filling phase are the three phases of plant

Table 2.7 Fixed Effect Regression Result of Irrigated and Rainfed Rice Yields

Variable	Model 1	Model 2
GDD	–0.008	–0.18**
	(0.017)	(0.070)
Rainfall	0.0014***	0.0008
	(0.0003)	(0.0005)
Linear trend	0.009	0.009
	(0.003)	(0.003)
Constant	7.97***	11.81***
	(0.47)	(1.69)
Observation	168	126
R²	0.69	0.62

Note: Dependent variable (Yield) is expressed in natural logarithm, and all exogenous variables are expressed in their physical form. Values in parentheses include robust standard errors. *Significant at 10%; **Significant at 5%; ***Significant at 1%.

growth. In case of rice, the temperature had a significant negative impact in vegetative and grain-filling phases. And the negative impact of temperature in case of maize and millet was found in the reproductive and grain-filling phases.

2.8 Mapping Districts in the Scale of Climate Change Vulnerability

In this section, we describe the key findings from the vulnerability assessment in identifying the regions vulnerable to climate change in Karnataka. Twenty districts of the state were considered to classify them according to the susceptibility to climate change in the agriculture sector with accounting for adaptation. Indicator based approach was employed to identify the climate change vulnerable regions. Vulnerability assessment was formulated as the function of three components namely, exposure, sensitivity and adaptive capacity. Exposure is defined as the degree to which agriculture productivity is affected, where variation in rainfall and temperature were used as an indicator. In sensitivity, four indicators that reflect the sensitivity of agriculture environment were employed. Agricultural labourers, cultivators, net sown area and rainfed area have been considered as indicators. When it comes to adaptive capacity several drivers that influence adaptation to climate change related effects were included in the study. Rural literacy rate, agriculture GDP, crop diversification, cropping intensity, gross cropped area, irrigated area and source of irrigation. In the current study, the assessment of climate change vulnerability in Karnataka has focused on performance instead of demographic indicators. The analysis was estimated by analysing the growth of appropriate vulnerability indicators instead of their cross-sectional data at various points of time in order to show the progress of vulnerability.

The indicators considered for the study were normalized in comparable range. Further, integration of the components to construct vulnerability index has been done with and without the weights. The present study assigns weights based on the factor analysis method. Separate factor analysis has been undertaken for exposure, sensitivity and adaptive capacity components using principal component analysis. Based on the variance the loadings from the first principal component were used as the weights. In case of the exposure component, the extremely vulnerable districts were from northern region (see Table 2.8 and Figure 2.2). In case of sensitivity, except Chitradurga district (southern district), other districts in north region were highly sensitive (see Table 2.9 and Figure 2.3). However, in case of adaptive capacity Raichur and Bijapura district were highly adaptive (see Table 2.10 and Figure 2.4). The integrated vulnerability assessment indicated that agriculture in the northern region is highly susceptible to climate change. The study showed that Gulbarga and Bidar districts were extremely vulnerable while Kodagu and Dakshin Kannada were categorized as low vulnerable. However, three districts namely, Chitradurga, Kolar and Tumkur districts from southern region were also highly susceptible to climate change (see Table 2.11 and Figure 2.5).

Table 2.8 District-Wise Exposure Component Scores and Degree of Exposure (with Weights)

District	RN 1	RN 2	Max 1	Max 2	Max 3	Max 4	Score	Degree
Bidar	0.040	0.259	0.854	0.206	0.372	0.899	0.928	Extreme
Gulbarga	0.049	0.096	0.783	0.191	0.366	0.798	0.805	Extreme
Raichur	0.073	0.096	0.482	0.140	0.366	0.675	0.645	Extreme
Bijapura	0.077	0.124	0.629	0.081	0.274	0.645	0.645	Extreme
Dharwad	0.209	0.111	0.275	0.051	0.145	0.593	0.488	Extreme
Chitradurga	0.048	0.068	0.349	0.078	0.180	0.534	0.444	High
Bellary	0.062	0.047	0.349	0.078	0.180	0.534	0.441	High
Shimoga	0.033	0.110	0.275	0.051	0.145	0.593	0.426	High
Belgaum	0.053	0.083	0.340	0.056	0.072	0.490	0.386	High
Kolar	0.033	0.052	0.206	0.150	0.283	0.266	0.349	High
Hassan	0.031	0.128	0.161	0.075	0.145	0.281	0.290	Moderate
Chikmangalur	0.033	0.160	0.267	0.053	0.009	0.292	0.287	Moderate
Bangalore U	0.047	0.106	0.206	0.125	0.202	0.102	0.278	Moderate
Bangalore R	0.028	0.000	0.057	0.150	0.283	0.266	0.277	Moderate
Tumkur	0.049	0.038	0.161	0.075	0.145	0.281	0.264	Moderate
U.Kannada	0.000	0.150	0.121	0.000	0.037	0.398	0.249	Low
D.Kannada	0.000	0.025	0.208	0.032	0.000	0.242	0.179	Low
Mysore	0.064	0.206	0.000	0.083	0.107	0.000	0.162	Low
Mandya	0.059	0.116	0.000	0.083	0.107	0.000	0.129	Low
Kodagu	0.024	0.058	0.038	0.031	0.040	0.031	0.078	Low

Note: RN 1, Kharif season rainfall; RN 2, Rabi season rainfall; Max 1, maximum temperature in post-rabi season; Max 2, maximum temperature in summer season; Max 3, maximum temperature in Kharif season; Max 4, maximum temperature in rabi season.

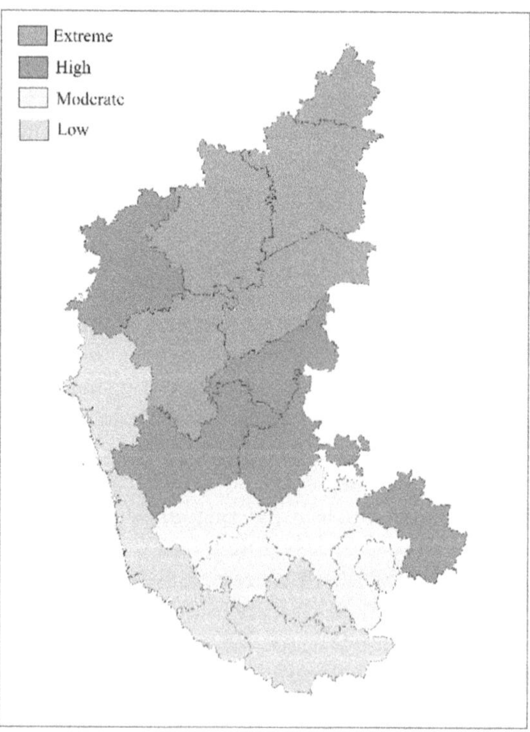

Figure 2.2 Classification of Districts Based on Exposure Component (with Weights).

Table 2.9 District-Wise Sensitivity Component Scores (with Weights)

Name	CV	LB	NSA	RA	Score	Degree
Gulbarga	0.165	0.786	0.719	0.109	0.877	Extreme
Bijapura	0.171	0.680	0.767	0.059	0.827	Extreme
Dharwad	0.171	0.722	0.640	0.109	0.810	Extreme
Raichur	0.171	0.765	0.538	0.056	0.755	Extreme
Belgaum	0.213	0.574	0.473	0.024	0.634	Extreme
Bellary	0.165	0.786	0.249	0.026	0.605	High
Chitradurga	0.220	0.680	0.224	0.100	0.604	High
Mysore	0.201	0.659	0.298	0.065	0.603	High
Tumkur	0.274	0.510	0.325	0.079	0.586	High
Shimoga	0.195	0.616	0.346	0.006	0.574	High
Bidar	0.128	0.659	0.189	0.115	0.538	Moderate
Kolar	0.213	0.595	0.181	0.082	0.529	Moderate
Hassan	0.323	0.425	0.199	0.085	0.510	Moderate
Mandya	0.287	0.531	0.108	0.000	0.457	Moderate
Bangalore Rural	0.244	0.425	0.144	0.085	0.444	Moderate
Chikmangalur	0.153	0.339	0.145	0.121	0.374	Low
Uttara Kannada	0.116	0.276	0.029	0.091	0.253	Low
Bangalore U	0.061	0.149	0.000	0.091	0.148	Low
Dakshin Kannada	0.024	0.064	0.107	0.044	0.118	Low
Kodagu	0.000	0.000	0.058	0.150	0.103	Low

Note: CV, agriculture cultivators; LB, agriculture labourers; NSA, net sown area; RA, rainfed area.

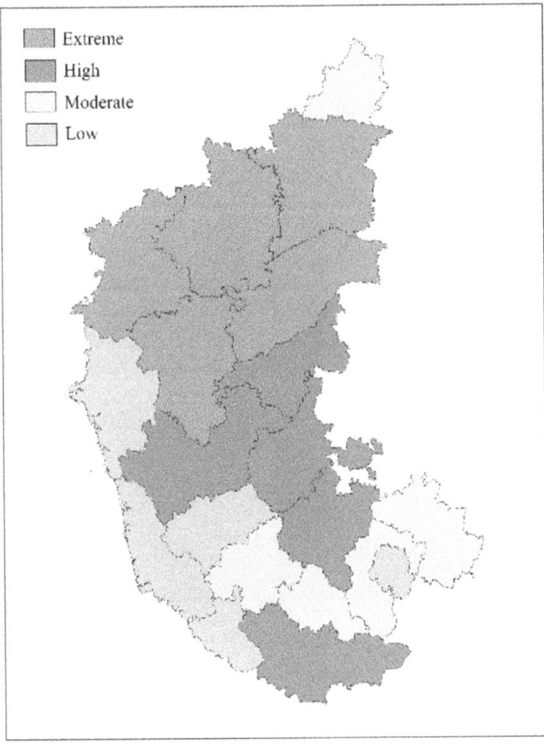

Figure 2.3 Classification of Districts Based on Sensitivity Component (with Weights).

Table 2.10 District-Wise Adaptive Capacity Component Score (with Weights)

Name	LR	GCA	CI	IL	CD	GDP	Canals	TBW	Score	Degree
Raichur	0.083	0	−0.005	0.037	0.001	0.063	1.001	0.999	0.651	Extreme
Bijapura	0.208	0	−0.012	0.061	0.001	0.044	0.755	0.754	0.541	Extreme
Shimoga	0.457	0.001	−0.001	0.011	0.001	0.043	0.539	0.538	0.475	Extreme
Mysore	0.187	0	0	0.01	0	0.032	0.569	0.568	0.409	Extreme
Mandya	0.333	−0.003	0.009	0.02	0	0.047	0.479	0.478	0.408	Extreme
Gulbarga	0	−0.001	−0.017	0.067	0.001	0.047	0.568	0.567	0.368	High
Belgaum	0.291	−0.002	0.029	0.077	0	0.037	0.349	0.349	0.338	High
Dharwad	0.395	−0.003	−0.022	0.067	0	0.022	0.32	0.32	0.329	High
Bellary	0.146	0	0.014	0.042	0.001	0.055	0.404	0.404	0.319	High
Hassan	0.457	−0.001	0.004	0.007	0.001	0.04	0.164	0.164	0.25	High
Dakshin Kannada	0.707	−0.001	−0.01	0.008	0	0.04	0	0	0.222	Moderate
Chikmangalur	0.519	0.001	0.008	0.091	0	0.047	0.029	0.029	0.217	Moderate
Uttara Kannada	0.582	0	0.004	0.074	0.002	0.059	0	0	0.216	Moderate
Bangalore Urban	0.519	−0.005	0.016	0.04	0.001	0.039	0	0.057	0.2	Moderate
Kodagu	0.623	0.003	0.017	−0.149	0.001	0.06	0.007	0.007	0.17	Moderate
Tumkur	0.415	−0.002	0.014	0.044	0.001	0.036	0.019	0.019	0.163	Low
Chitradurga	0.374	−0.001	−0.002	0.046	0.001	0.044	0.008	0.008	0.143	Low
Bidar	0.249	0	0.01	0.044	0.001	0.072	0.03	0.03	0.13	Low
Bangalore Rural	0.374	−0.002	−0.009	−0.004	0.001	0.032	0.01	0.01	0.123	Low
Kolar	0.312	0.002	0.047	−0.018	0	0.048	0	0	0.117	Low

Note: LR, literacy rate; GCA, gross cropped area; CI, cropping intensity; IL, irrigated area; CD, crop diversification; GDP, gross domestic product; Canals, canal irrigation; TBW, tube well and bore well irrigation.

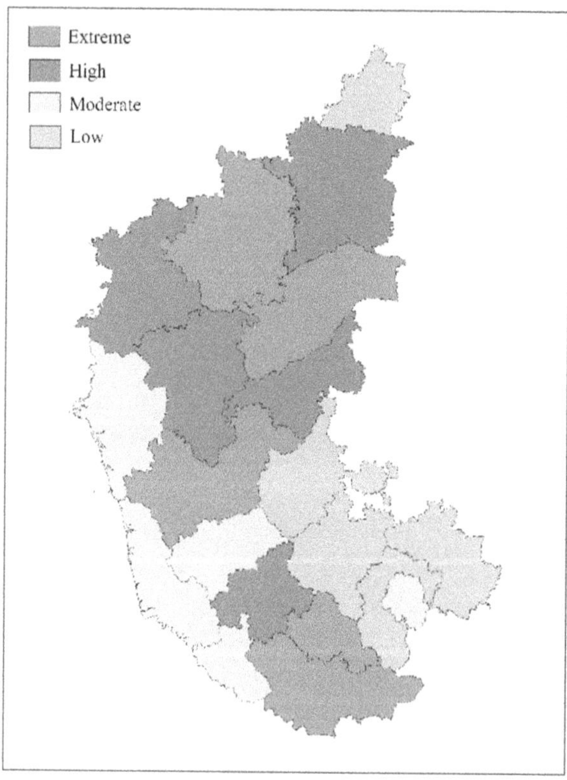

Figure 2.4 Classification of Districts Based on Adaptive Capacity Component (with Weights).

Table 2.11 District-Wise Vulnerability Index Score (with Weights)

District	EX	SI	AC	VS	Degree
Bidar	0.93	0.54	0.13	1.34	Extreme
Gulbarga	0.80	0.88	0.37	1.31	Extreme
Dharwad	0.49	0.81	0.33	0.97	Extreme
Bijapura	0.65	0.83	0.54	0.93	Extreme
Chitradurga	0.44	0.60	0.14	0.90	Extreme
Kolar	0.35	0.53	0.12	0.76	High
Raichur	0.65	0.75	0.65	0.75	High
Bellary	0.44	0.60	0.32	0.73	High
Tumkur	0.26	0.59	0.16	0.69	High
Belgaum	0.39	0.63	0.34	0.68	High
Bangalore R	0.28	0.44	0.12	0.60	Moderate
Hassan	0.29	0.51	0.25	0.55	Moderate
Shimoga	0.43	0.57	0.48	0.52	Moderate
Chikmangalur	0.29	0.37	0.22	0.44	Moderate
Mysore	0.16	0.60	0.41	0.36	Moderate
U.Kannada	0.25	0.25	0.22	0.29	Low
Bangalore U	0.28	0.15	0.20	0.23	Low
Mandya	0.13	0.46	0.41	0.18	Low
D.Kannada	0.18	0.12	0.22	0.07	Low
Kodagu	0.08	0.10	0.17	0.01	Low

Note: EX, exposure component; SI, sensitivity component; AC, adaptive capacity component; VS, vulnerability index.

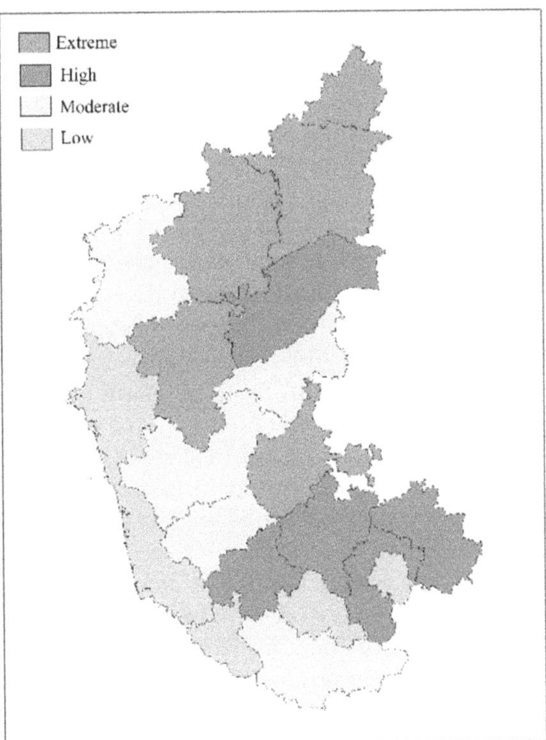

Figure 2.5 Classification of Districts Based on Vulnerability Index (with Weights).

2.9 Adaptation Measures Essential to Deal with Climate Change

The results from the current study indicate that agriculture in Karnataka is highly vulnerable to climate change, which needs special attention from policymakers to reduce the negative effect of climate change. Kharif season temperature and rainfall had significant impact on the net revenue among all seasonal climate variables. Temperature had a negative effect on revenue, while rainfall had a positive influence. The productivity of rice, maize and finger millet declined by 5%, 6%, and 16%, respectively, due to climate change impact for the period 1992–2012. The estimated damages for rainfed and cultivation are larger and significant. Results show a decline of yields in the order of 13% and 17% for maize and rice, respectively, which draws serious concern. The findings from the current study have clearly established the threat of climate change to agricultural production systems in southern India. What seems to be concerning is the extent of crop loss observed due to temperature warming given that these estimated effects are robust to different alternative climate scenarios. The study has identified climate related vulnerable regions in Karnataka. North interior region of Karnataka is highly susceptible to climate change when compared to south interior and coastal regions.

From the above findings the study suggests a number of policy implications pertaining to the specific research issues addressed. First, considering the negative impact of climate change on agricultural net revenue, suitable adaptation measures have to be developed. Several household survey-based studies have observed that although farmers do realize a steady change in regular climate patterns over the years, they do not consider this change as a reason for any action to be taken in adjusting their cropping patterns or farm management practices (Chaudhary and Bawa, 2011; Budhakoti and Zander, 2019). Both the central government as well as state government have climate policies in place that consider agriculture as a prime focus, however specific technical knowledge and technical support need to be made available to farmers for informative climate resilience action. Diversifying to non-farm income sources should be encouraged among vulnerable communities to reduce sole dependence on agriculture for livelihood. Secondly, the current study has observed that rise in temperature and shortage of rainfall have reduced the crop yield for several cereal crops. This has grave concerns for food security in general. There is a need for technology-based solution to agricultural woes in India. Sensor-based motorized instruments can determine the precise amount of irrigation, fertilizer application, and other input used to achieve optimal yield. These instruments may seem expensive in the beginning for farmers but the long-term gains are significant. There is a role for government to financially support the spread of such technology-based agricultural practices termed as "precision agriculture" or "climate smart agriculture". Labour shortage has been observed in farms in many of developing countries. Technological solutions can replace traditional farming practices to make agriculture more climate resilient.

Third, there is a need for strategic farming in which the crops are chosen according to climate suitability. Crops like millets or sorghum are suitable in dry arid and semi-arid climates, but farmers still engage in rice cultivation in these areas because market price for rice is much higher than the coarse cereals. These market dynamics create unwarranted outcomes wherein farmers permanently shift to rice cultivation in search of higher revenue but eventually fall in a vicious trap when the crop yield is damaged due to adverse weather conditions. Government can intervene by increasing the minimum support price for crops like millet to encourage farmers to cultivate them again. The vulnerability assessment in the current study has indicated that the NIK region is highly susceptible to climate change, where special attention to adaptive measures like provision of heat tolerant seeds at subsidized price and suitability of crop choice based on the climate information has to be suggested to the farmers. Furthermore, the literacy rate in this region is low when compared to other two regions. Institutional support that focuses on climate change adaption has to be set up to educate the farmers to reduce the climate shocks in future. While in SIK and CK region the farmers should focus on adopting modern techniques and better irrigation facilities to improve the production and to withstand climate change. Further, farmers need to adapt a significant shift in crop cultivation from high water-demanding crops to heat-resistant crops such as millet to meet the needs of growing population in the presently changing climate scenario.

Finally, farmer organizations in the form of cooperatives can also help farmers to organize themselves in an inclusive decision-making body to support each other. This should act as a "bottom up" approach in which farmers take key decisions through consensus or majority principle. Such farmer organizations also make it easy for the government and external funding agencies to support smallholder farmers. For example, Fairtrade certification that aims to support smallholder farmers through a minimum "floor price" at the time of market price distress requires member farmers to be grouped under a formal cooperative.

Notes

1 Disclaimer – This chapter is a revised version of one of the chapters from a Ph.D. thesis supervised by one of the authors of this book manuscript. Thesis title – Assessing the Impact of Climate Change on Agriculture: Empirical Evidence from a South Indian State. Ph.D. Scholar – Kalli, Rajesh M. Ph.D. Supervisor – Jena, Pradyot Ranjan
2 Desiccation is the state of extreme dryness, or the process of extreme drying.

References

Abeysingha, N. S., Singh, M., Islam, A. and Sehgal, V. K. (2016) 'Climate change impacts on irrigated rice and wheat production in Gomti River basin of India: A case study', *Springer Plus*, 5, pp. 1–20.

Adams, R. M., Rosenzweig, C., Peart, R. M., Ritchie, J. T., McCarl, B. A., Glyer, J. D., Curry, R. B., Jones, J. W., Boote, K. J. and Allen, L. H. (1990) 'Global climate change and US agriculture', *Nature*, 345(6272), pp. 219–224. https://doi.org/10.1201/9781420077544.ch16

Adger, W. N., Huq, S., Brown, K., Declan, C. and Mike, H. (2003) 'Adaptation to climate change in the developing world', *Progress in Development Studies*, 3(3), pp. 179–195.

Alam, M. J., Ahmed, K. S., Sultana, A., Firoj, S. M. and Hasan, I. M. (2018) 'Ensure food security of Bangladesh: Analysis of post-harvest losses of maize and its pest management in stored condition', *Journal of Agricultural Engineering and Food Technology*, 5(1), pp. 26–32.

Alexandratos, N. and Bruinsma, J. (2003) 'World agriculture: Towards 2015/2030: An FAO perspective', *Land Use Policy*, 20(4), p. 375.

Antle, J. M. (1995) 'Climate change and agriculture in developing countries', *American Journal of Agricultural Economics*, 77(3), pp. 741–746.

Anwar, M. R., Liu, D. L., Macadam, I. and Kelly, G. (2013) 'Adapting agriculture to climate change: A review', *Theoretical and Applied Climatology*, 113(1–2), pp. 225–245.

Bandara, J. S. and Cai, Y. (2014) 'The impact of climate change on food crop productivity, food prices and food security in South Asia', *Economic Analysis and Policy*, 44(4), pp. 451–465.

Basha, G., Kishore, P., Ratnam, M. V., Jayaraman, A., Agha Kouchak, A., Ouarda, T. B. and Velicogna, I. (2017) 'Historical and projected surface temperature over India during the 20th and 21st century', *Scientific Reports*, 7(1), p. 2987.

Bhende, G., & Tembhare, G. (2013) 'Stress intensification & flexibility in pipe stress analysis', *International Journal of Modern Engineering Research*, 3(3), pp. 1324–1329.

Burney, J. and Ramanathan, V. (2014) 'Recent climate and air pollution impacts on Indian agriculture', *Proceedings of the National Academy of Sciences*, 111(46), pp. 16319–16324.

Carleton, T. A. (2017) 'Crop-damaging temperatures increase suicide rates in India', *Proceedings of the National Academy of Sciences*, 114(33), pp. 8746–8751.

Chaudhary, P. and Bawa, K. S. (2011) 'Local perceptions of climate change validated by scientific evidence in the himalayas', *Biology Letters*, 7(5), pp. 767–770.

FAO. (2002) 'World agriculture: Towards 2015/2030,' *Food and Agriculture Organization*, 20(4), pp. 97.

FAO. (2010) 'Climate-smart agriculture: Policies, practices and financing for food security, adaptation and mitigation', *FAO*, 49.

Guhathakurta, P., Rajeevan, M., Sikka, D. R. and Tyagi, A. (2015) 'Observed changes in southwest monsoon rainfall over India during 1901-2011', *International Journal of Climatology*, 35, pp. 1881–1898.

Guiteras, R. (2009) The impact of climate change on Indian agriculture. Manuscript, Department of Economics, University of Maryland, College Park, Maryland.

Gupta, R., Somanathan, E. and Dey, S. (2017) 'Global warming and local air pollution have reduced wheat yields in India', *Climatic Change*, 140(3-4), pp. 593–604.

Hitz, S. and Smith, J. (2004) 'Estimating global impacts from climate change', *Global Environmental Change*, 14(3), pp. 201–218. https://doi.org/10.1016/j.gloenvcha.2004.04.010

IPCC. (2007) 'Climate models and their evaluation.' *Climatic Change* 2007 Phys. Sci. Basis.

Jayaraman, T. and Murari, K.. (2014) 'Climate change and agriculture: Current and future trends, and implications for India', *Review of Agrarian Studies*, 4(1).

Jena, P. R. and Grote, U. (2022) 'Do certification schemes enhance coffee yields and household income? Lessons learned across continents,' *Frontiers in Sustainable Food Systems*, 5, p. 716904.

Kaiser, H. M., Riha, S. J., Wilks, D. S., Rossiter, D. G. and Sampath, R. (1993) 'A farm-level analysis of economic and agronomic impacts of gradual climate warming', *American Journal of Agricultural Economics*, 75(2), pp. 387–398. https://doi.org/10.2307/1242923

Kalli, R. and Jena, P. R. (2022a) 'Nd climate indicators to classify vulnerable regions in the IndiCombining agriculture, social aan semi-arid region', *Journal of Water and Climate Change*, 13(2), pp. 542–556. https://doi.org/10.2166/wcc.2021.197

Kumar, K. S. K. and Parikh, J. (2001) 'Socio-economic impacts of climate change on Indian agriculture', *International Review for Environmental Strategies*, 2(2), pp. 277–293.

Kumar, K. S. K. and Parikh, J. (2001) 'Indian agriculture and climate sensitivity', *Global Environmental Change*, 11(2), pp. 147–154.

Lobell, D. B., Sibley, A. and Ivan Ortiz-Monasterio, J. (2012) 'Extreme heat effects on wheat senescence in India', *Nature Climate Change*, 2(3), pp. 186–189. https://doi.org/10.1038/nclimate1356

Mall, R. K., Singh, R., Gupta, A., Srinivasan, G. and Rathore, L. S. (2006) 'Impact of climate change on Indian agriculture: A review', *Climatic Change*, 78, pp. 445–478.

Mendelsohn, R. (2008) 'The impact of climate change on agriculture in developing countries', *Journal of Natural Resources Policy Research*, 1(1), pp. 5–19.

Mendelsohn, R. and Tiwari, D. (2000) *Two essays on climate change and agriculture: A developing country perspective* (No. 145).

Mirza, N., Pervez, A., Mahmood, Q., Shah, M. M. and Shafqat, M. N. (2011) 'Ecological restoration of arsenic contaminated soil by arundo donax l', *Ecological Engineering*, 37(12), pp. 1949–1956. https://doi.org/10.1016/j.ecoleng.2011.07.006

Mishra, D., Sahu, N. C. and Sahoo, D. (2016) 'Impact of climate change on agricultural production of Odisha (India): A Ricardian analysis,' *Regional Environmental Change*, 16, pp. 575–584.

Nelson, E., Mendoza, G., Regetz, J., Polasky, S., Tallis, H., Cameron, D. R., Chan, K. M. A., Daily, G. C., Goldstein, J., Kareiva, P. M., Lonsdorf, E., Naidoo, R., Ricketts, T. H. and Shaw, M. R. (2009) 'Modeling multiple ecosystem services, biodiversity conservation, commodity production, and tradeoffs at landscape scales', *Frontiers in Ecology and the Environment*, 7(1), pp. 4–11. https://doi.org/10.1890/080023

Nordhaus, W. D. (1993) 'Optimal Greenhouse-Gas Reductions and Tax Policy in the "DICE" Model Author (s): William D. Nordhaus Source : The American Economic Review, May, 1993, Vol. 83, No. 2', *Papers and Proceedings of the Hundred and Fifth Annual Meeting of the Americ.* 83(2), pp. 313–317.

Revadekar, J. V., Kothawale, D. R., Patwardhan, S. K., Pant, G. B. and Rupa Kumar, K. (2012) 'About the observed and future changes in temperature extremes over India', *Natural Hazards*, 60(3), pp. 1133–1155.

Rosenzweig, C. and Hillel, D. (2008) *Climate variability and the global harvest: Impacts of El Niño and other oscillations on agro-ecosystems.* Oxford University Press.

Rosenzweig, C., Iglesias, A., Yang, X. B., Epstein, P. R. and Chivian, E. (2001) ''Climate change and extreme weather events - implications for climate change and extreme weather events – implications for food production, plant diseases, and pests food production, plant diseases, and pests', *Global Change and Human Health*, 2(2), pp. 90–104. http://link.springer.com/10.1023/A:1015086831467

Rosenzweig, C. and Parry, M. L. (1994) 'Potential impact of climate change on world food supply', *Nature*, 367(6459), p. 133.

Varadan, R. J., Kumar, P., Jha, G. K., Pal, S. and Singh, R. (2015) 'An exploratory study on occurrence and impact of climate change on agriculture in Tamil Nadu, India', *Theoretical and Applied Climatology*, 127(3–4), pp. 993–1010. https://doi.org/10.1007/s00704-015-1682-9

Yoshida, S. (1981) 'Climatic environment and its influence', *Fundamentals of Rice Crop Science*, 65–109. http://books.irri.org/9711040522_content.pdf

3 An Overview of Government Policies and Strategies on Climate Smart Agriculture in India

3.1 Introduction

The Food and Agriculture Organization (FAO) coined a new term, "Climate-Smart Agriculture", at The Hague Conference on Agriculture, Food Security and Climate Change in 2010. FAO has introduced the climate-smart agriculture (CSA) concept to address three objectives "to increase productivity in agriculture, promote adaption to climate change and to mitigate the climate change". National food security and development goals will be achieved through CSA practices. The CSA approach guides the implementation of environmentally friendly and resilient practices in agri-food systems. FAO (2013) advocates that by adopting CSA, it is possible to work towards achieving global goals such as the SDGs and the Paris Agreement, which focuses on addressing climate change and promoting sustainability. FAO (2010) has defined the three broad objectives of CSA

1 To increase crop productivity, farm income, and food security.
2 To adopt climate-resilient agricultural practices at multiple levels of agricultural production.
3 To mitigate the environment by reducing greenhouse gas emissions from the agriculture sector (Lipper et al., 2010).

FAO has produced a sourcebook that provides the most extensive advice on climate-smart agriculture. This sourcebook includes a range of practical strategies and techniques that form the basis of CSA (FAO, 2013). The sourcebook elaborates on the concept of CSA and demonstrates its potential, makes it uniform, and has limitations. The sourcebook encompasses various modules that look into topics such as climate-smart agriculture, forestry, and fisheries, landscape management for climate-smart agricultural ecosystems, water management, soil management for CSA, energy, conservation, and sustainable use of genetic resources for food and agriculture. Additionally, it covers crop production systems, climate-smart forestry, climate-smart fisheries and aquaculture, sustainable and inclusive food value chains, financing climate-smart agriculture, and disaster risk reduction.

In recent times, there has been a significant increase in the attention given to CSA. The various stakeholders on this earth, such as international organizations,

DOI: 10.4324/9781003290445-3

national governments, farmers, civil society organizations (CSOs), the private sector, the research community and at the root level, the farmers, have taken steps to implement CSA initiatives (Dinesh et al., 2015). The farmers, researchers, civil society, private sectors, and policymakers coordinate to promote CSA through four main action areas: (1) identifying the core issues of climate change; (2) promoting climate and agriculture policy convergence; (3) increasing the local institution's activeness; and (4) interlink the climate and agriculture financing. CSA doesn't follow "business-as-usual" approaches; it follows flexible, context-specific solutions backed by suitable innovative policy and financing actions (Lipper et al., 2014). The author is concerned that adopting CSA by the most vulnerable groups, including smallholder producers and poor and marginalized communities, should be prioritized by identifying the barriers to adoption. There is a need for appropriate policies and planning to be developed that take into account the specific agroecological production systems present in different regions.

Dinesh et al. (2015) reviewed and analysed 19 case studies on climate-smart agriculture. The author found that all of them have contributed towards the sustainable production of food, increased productivity, enhancement of food security, and increased farm incomes and development. The other co-benefits of adopting CSA are employment generation, health and nutritional benefits, and infrastructure development. The studies also showed that CSA adoption positively impacts gender and social inequalities. Campbell et al. (2014) have connected CSA and Sustainable Intensification (SI) and found that these two practices complement each other. SI and CSA are closely interlinked practices where both emphasize on achieving the adaptation and mitigation goal.

FAO has advocated the Landscape approach to the CSA. According to FAO (2012), "Landscape approach refers to a set of concepts, tools, methods and approaches deployed in landscapes in a bid to achieve multiple economic, social, environmental objectives through processes that recognize, reconcile and synergize interests, attitudes and actions of multiple actors". Adopting a landscape approach incorporating land-use planning makes it possible to mitigate conflicts arising from resource utilization and tackle the dangers facing biodiversity-rich ecosystems such as forest areas and wetlands (Scherr et al., 2012). Furthermore, implementing such an approach can aid in restoring crucial ecosystem functions and services. This method has the potential to produce advantageous and lasting implications for populations encountering unpredictable and extreme weather events (IPCC, 2014; FAO, 2015).

FAO has prescribed a set of climate-smart crop production practices and technologies which focuses on the adoption of specific climatic hazards and practices that simultaneously reduce production risks and vulnerabilities (FAO, 2017). The practices include genetically modified improved seed varieties which tolerate extreme conditions of droughts and floods. The other practices are integrated pest management, improved water use and management, sustainable soil and land management, and sustainable mechanization, which enable to face climate distress. In particular, FAO has prescribed the following CSA adoptions, include the use of stress-tolerant seeds, crop diversification, crop rotation, crop residues management, soil mulching, agroforestry, integrated nutrient management, rainwater harvesting, in-situ water conservation, minimizing mechanical soil disturbance or no-tillage, sustainable agricultural mechanization, integrated crop-livestock systems, deficit irrigation, precision water applications, high-efficiency pumps and improving drainage (FAO, 2013; FAO, 2017). Disaster risk reduction measures have the potential to aid CSA in achieving its goals, especially in terms of enhancing the adoption to climate change and strengthening the capacity of agricultural communities and ecosystems to cope with climate variations and shifts. Effective disaster risk reduction policies, programmes, and practices can be useful tools to promote and expand CSA (Mitchell et al., 2010). Azadi et al. (2021) proposed a five-component framework that includes farmers predicting key incidents, quantifying the impact of incidents, identifying farmers' coping methods, assessing farmers' livelihood resources during a crisis, and adopting to climate incidents.

3.2 Adaptation vs Adoption

IPCC (2001) states, "Adaptation is the adjustment process to actual or expected climate and its effects. Adoption seeks to moderate or avoid harm or exploit beneficial opportunities in human systems. In some natural systems, human intervention may facilitate adjustment to expected climate and its effects" (IPCC, 2001). Adger et al. (2005) define "Adaptation as anything that reduces the risks associated with climate change, and vulnerability to climate change impacts, in both the short- and long-term, for both the direct beneficiary of the adoption and the wider society without compromising economic, social, and environmental sustainability".

According to the Nhemachena and Hassan, (2007), "Adaptation is the process of improving society's ability to cope with changes in climatic conditions across time scales, from short term (e.g., seasonal to annual) to the long term (e.g., decades to centuries)". Zilberman et al. (2012) state, "Adaptation is a change in practice or technology used by economic agents or a community, has a long intellectual history". Adoption is typically assessed through a discrete choice analysis, where individuals are presented with options and must choose one. This analysis may also involve a continuous indicator measuring the degree or extent of adoption. For example, in a study of new technology adoption, the discrete choice might involve selecting whether or not to adopt the technology. At the same time, the continuous indicator could measure how much the individual uses the technology once they have adopted it (Zilberman et al., 2012).

Katz Elihu (1961) divided the adoption decision into five stages: awareness, interest, evaluation, trial, and adoption. These decisions are influenced by the learning, understanding and judgement associated with adoption. Roger's innovation-decision model passes through five stages, shown in Figure 3.1. Knowledge, persuasion, decision, implementation, and confirmation are examples of these processes. Individuals in the knowledge stage are exposed to new concepts and develop understanding. In the persuasion stage, the individual either persuades others or is open to persuasion. In the decision stage, the individual

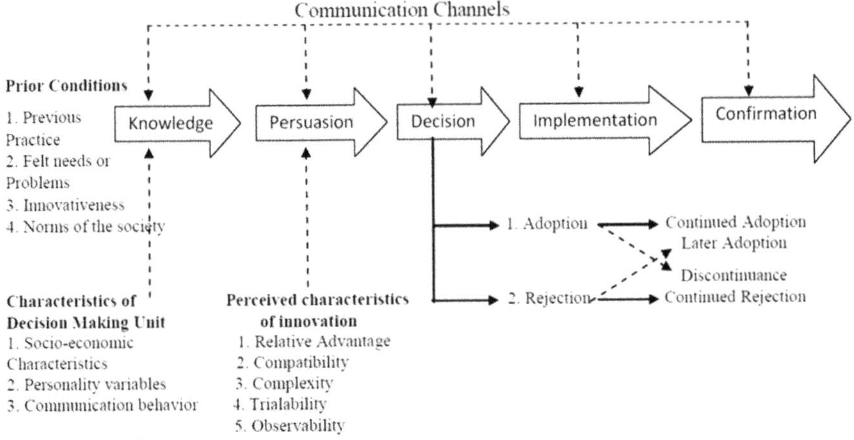

Figure 3.1 Roger's Innovation-Decision Model.

Source: Rogers (1983).

decides whether to adopt or reject the new idea. In the implementation stage, the individual implements the decision made in the previous stage. In the confirmation stage, the individual continues questioning the knowledge of the decision. A growing body of research on adoption signifies that adopting various conservation agriculture practices helps build a climate-resilient system. According to Knowler and Bradshaw (2007), adopting low or no-tillage practises improves soil conditions and sequesters carbon in both developed and developing countries. Adopting water-saving technologies, such as drip irrigation, can enhance the effectiveness of resource usage and potentially result in increased crop yield. This approach may also result in lower water intensity in some cases and consistently reduces drainage (Schoengold and Zilberman, 2007). Access to financial incentives has a significantly positive effect on adopting conservation practices (Linn, 2008). Adoption can be distinguished between micro and macro levels. Micro adoption involves selecting discrete strategies, such as adopting existing technologies, migration, or changes in input use. Macro adoption is measured by aggregate behaviour and involves policy rule changes at the village, country, or global level. The literature on adaptation vs adoption emphasizes decisions taken by the farmers for a particular practice or technology. "Adoption" focuses on new technologies, whereas "adoption" focuses on existing ones (Feder et al., 1985).

3.3 Climate Smart Agriculture in India

The Indian government launched the National Action Plan on Climate Change (NAPCC) in 2008 with the goal of addressing climate change challenges on a national scale while ensuring the country's continuing development. The strategy is divided into eight major missions that focus on minimizing and adapting to the consequences of climate change in various fields. The National Adoption Fund for Climate Change (NAFCC) was established in August 2015 to assist Indian states and union territories that are particularly vulnerable to the detrimental effects of climate change. Its major goal is to finance the costs of climate change adaptation through the National Implementing Entity (NIE). NABARD (National Bank for Agriculture and Rural Development) is the National Implementing Entity (NIE) to undertake adoption initiatives under the National Adoption Fund for Climate Change (NAFCC). In 2014–2015, the National Mission for Sustainable Agriculture (NMSA) was launched to enhance agricultural productivity, sustainability, profitability, and climate resilience. Improving farming practices, improving the variety of seeds, animal and fish culture, water usage efficiency, pest management, agricultural insurance, nutrient management, credit support, access to markets, access to information, and livelihood diversification are all significant adaptation approaches undertaken by NMSA.

Consultative Group on International Agricultural Research (CGIAR), a Research Programme on Climate Change, Agriculture, and Food Security (CCAFS), has been partnering with national programmes to collaborate with rural communities in developing Climate-Smart Villages (CSVs). These villages are examples of local initiatives that ensure food security, encourage adoption and enhance resilience against climate-related challenges (Table 3.1).

Anwar et al. (2013) found that extreme weather events and disasters adversely affect the agricultural sector of developing countries. The impact of climate change on agriculture is unpredictable and ultimately depends on various climatic factors and extreme occurrences, including droughts and floods. The impact of climate change has been categorized into two aspects: the bio-physical aspect and the socioeconomic aspect (Anwar et al., 2013). As this study does not primarily focus on the impact of climate change on agriculture, the literature review has been done on the general impact of climate change on agriculture

Table 3.1 Climate-Smart Village Approach by CCAFS

	CSA approach	*Adoption practices*
1	Weather-smart activities	ICT-based agro-advisories, stress-tolerant crops, index-based insurance, weather forecasts
2	Water-smart practices	Crop diversification, laser land levelling, water conservation, resilient water management practices, direct-seeded rice, rainwater harvesting, drip irrigation, raised bed planting, alternate wetting and drying in rice.
3	Carbon-smart practices	Diversified land-use systems, conservation tillage, Agroforestry, livestock and manure management, and residue management
4	Nitrogen-smart practices	Site-specific nutrient management, residue management and legume catch-cropping, leaf-colour charts, precision fertilizer application, hand-held crop sensors
5	Energy-smart technologies	Biogas systems, fuel-efficient agricultural machinery, minimum tillage, solar energy irrigation
6	Knowledge-smart activities	Enhancement of CSA capacity, cross-site farmer visits, farmer-to-farmer learning, seed packets of adopted varieties, market and off-farm risk management system, and community seed and fodder banks

Source: Ghosh (2019).

around the globe and in India. The impact of climate change literature shows two broad types of impact assessment models: "general equilibrium models" and "partial equilibrium models". However, this literature survey will be reported chronologically on the historical climate change impact studies. Adams et al. (1990) conducted a study to investigate the potential impact of climate change and increased CO_2 on agriculture in the US. Using the predictions of global climate models (GCMs), they found that changes in temperature and precipitation could lead to decreased yields of wheat, maize, and soybean and increased water requirements for crops. Nordhaus (1993) estimated the economic effect of doubling atmospheric carbon dioxide and climate change on world agriculture. Changes in domestic yields affect the prices of agricultural commodities, and changes in global consumption and production affect economic well-being. Kaiser et al. (1993) analysed the potential economic and agronomic effects of climate change. Moderate warming had a positive impact on crop yield; extreme warming had a negative impact on yield. The increase in global mean temperature (GMT) and impacts were not consistent across sectors. Some sectors exhibited increasing adverse impacts with increasing GMT, particularly coastal resources, biodiversity, and possibly marine ecosystem productivity (Hitz and Smith, 2004). Rosenzweig and Hillel (1995) examined the effects of climate change on global food production and agriculture. Physical impacts of climate change could negatively impact crop yields, pest and disease pressure, and soil fertility (Antle, 1995). The physical impact of climate change could lead to the hindrance of economic development of the region. Climate change has adverse economic and social impacts, including food security concerns and decreased agricultural productivity (Mendelsohn and Tiwari, 2000).

Rosenzweig et al. (2001) found that changes in temperature, precipitation, and extreme weather events are likely to have significant implications for food production and the prevalence and severity of plant diseases and pest outbreaks in developing countries. Mendelsohn (2008), in an insightful review of the potential effects of climate change on agricultural production in developing countries, concluded that tropical and subtropical agriculture in developing countries is more climate-sensitive than temperate agriculture. Kumar and Parikh (2001) examined the socioeconomic impact of climate change on Indian agriculture. Climate change-induced yield shocks on India's gross domestic product (GDP) could result in a decline of 1.8–3.4%, with the reduction in agricultural GDP being the major contributing factor. Kumar and Parikh (2001) examined the sensitivity of Indian agriculture to these physical impacts. A 2°C increase in temperature and a 7% increase in precipitation could result in an estimated loss of about 8.4% of the total net revenue. The study by Guiteras (2009) assessed the impact of stochastic inter-annual weather fluctuations on agricultural productivity over 40 years. The study revealed that the anticipated climate change during the timeframe of 2010–2039 is expected to result in a decrease of 4.5–9% in the yields of significant crops. Nelson et al. (2009) used a global economic model, a crop simulation model, and hydrological models to assess climate change. They projected that, by 2050, the yields of major crops in developing countries could decline by as much as 10%, leading to significant food price increases and reductions in food security.

Lobell et al. (2012) assessed the impact of extreme heat on wheat production by using nine years of satellite measurements of wheat grown in northern India and found that temperatures during the grain-filling stage of wheat growth exceeded a certain threshold and that the wheat plants died prematurely, resulting in lower yields. Gupta et al. (2017) examined the impact of climate change on wheat production in India. Rising temperatures and air pollution have contributed to a decline in wheat yields in India. Higher temperatures during the wheat growing season have reduced yields, with the negative impact being more pronounced in regions with high levels of air pollution. Abeysingha et al. (2016) examined the impact of climate change on rice and wheat production in the Gomti river basin of India. The study found that the temperatures and changes in precipitation patterns could lead to lower crop yields, with the negative impact being more pronounced for rice than wheat. Mishra et al. (2016) investigated the impact of climate change on agricultural production in Odisha, a state in eastern India. The Ricardian analysis shows higher temperatures during the growing season and increased rainfall variability are associated with lower farm-level net revenue of paddy. Bangladesh, India, and Pakistan are experiencing more frequent and severe flooding between 0 and 2°C GMT changes (Mirza, 2011). Bandara and Cai (2014) examined the potential impacts of climate change on food crop productivity, food prices, and food security in South Asia, namely Bangladesh, India, Nepal, Pakistan, and Sri Lanka. Tim Wheeler (2013) reviewed the aspects of climate change and food security. The effect of climate change on crop productivity could have a negative impact on progress towards ending world hunger, with potential consequences for food availability.

3.4   The Empirical Literature on the Adoption of CSA Practices and Its Determinants

According to Smit et al. (1999), "adoption" is a process that pertains to all climate-sensitive fields, including forestry, agriculture, water management, coastal protection, public health, and disaster mitigation. Adoption is modifying something or altering one's conduct to suit a new purpose or circumstance. Adoption of climate change entails many measures to lessen susceptibility to diverse climatic extremes. Adoption of climate change is highly context-specific because it is contingent on the target region's climatic, environmental, social, and political conditions and industry. Fankhauser et al. (1999) have discussed various adoption strategies to address climate change. They suggested long-term weather-sensitive capital investments, sustainable development practices, coastal development plans, and drought contingency plans that should be revised to incorporate climate change considerations. CSA refers to a set of agricultural practices that promote food security in the context of climate change. This strategy incorporates climate change into sustainable agricultural practises to make agriculture more resilient to climate variability and lessen its contribution to global warming (FAO, 2010). The following group of studies in Table 3.2 shows the adoption of CSA practices around the globe and in India. By reviewing empirical papers on

Table 3.2 Adoption of CSA Practices Over the Years

Studies	Adoption strategies	Countries
Deressa et al. (2009)	Agroforestry, soil conservation, changing crop varieties, changing planting dates, and changing irrigation methods.	Ethiopia
Bryan et al. (2009)	Change in crop varieties, agroforestry, soil conservation, changing planting dates, and irrigation.	Ethiopia
Fosu-Mensah et al. (2012)	Crop diversification, planting of short-duration varieties seeds, changes in crop species, reducing farm size, change in planting date, and finding off-farm jobs.	Ghana
Jim et al. (2012)	Hybrid seeds, High Yielding Variety seeds, row planting and seedling, levelling (land preparation), leaf colour chart (LCC), shallow tube wells, surface water pumps, postharvest management, and farm mechanization.	Philippines
Tambo and Abdoulaye (2013)	Mixed cropping, early maturity varieties seeds, change in crop varieties, change in planting dates, water conservation, shift to non-farm work, planting of trees/shading for animals.	Nigeria
Abdur Rashid Sarker et al. (2013)	Direct seeded rice, short-duration rice varieties, change in planting dates, change in harvesting dates, agroforestry, change in crop varieties, cultivation of nitrogen enabled crops such as pulses.	Bangladesh
Bryan et al. (2013)	Rescheduling planting dates, change in crop type, crop diversifying, supplementing livestock feeds, change in fertilizer application, soil and water conservation practices	Kenya
Tessema et al. (2013)	Early planting, tree planting, terracing, micro irrigation, and water harvesting.	Ethiopia
Abid et al. (2015)	Change in crop variety, rescheduling planting dates, agroforestry, soil conservation, change in fertilizer, irrigation, and crop diversification.	Pakistan
Belay et al. (2017)	Crop diversification, planting date adjustment, soil and water conservation, integrating crops with livestock, and agroforestry.	Ethiopia
Elum et al. (2017)	DRS seeds, integrated pest management, change in planting date, and diversified and relocated crop insurance.	South Africa
Tripathi and Mishra (2017)	Crop diversification, agroforestry, and increasing use of groundwater for irrigation.	India
Alam et al. (2017)	Change in planting time, crop rotation, paddy-pulses, agroforestry, cultivating HYV rice varieties, and integrated farming.	Bangladesh
Kumar et al. (2018)	Maize diversification, direct seeded rice, aerobic rice cultivation, zero tillage, early harvest, and increased soil carbon (carbon sequestration).	India
Swami and Parthasarathy (2020)	Crop diversification, drought-resistant crops (DRC), change in planting/ harvesting date (PHD), drip irrigation, and crop insurance.	India
Singh (2020)	Improved irrigation facilities, change in cropping pattern, switch to non-farm occupation, use of early maturing varieties, and use of less water-consuming crops.	India

(Continued)

Table 3.2 (Continued)

Studies	Adoption strategies	Countries
Funk et al. (2020)	Information and communication technology (ICT), crop diversification, improved irrigation, and integrated farming.	India
Bahinipati and Venkatachalam (2015)	Crop diversification, changes in crop varieties, altering crop calendar and land holidays, soil conservation, pest and diseases management.	India
Ward and Makhija (2018)	Change in crop varieties.	India
Singh et al. (2018b)	Water conservation techniques, use of crop varieties of suitable duration, crop insurance and participation in non-farm activities.	India
Singh et al. (2018a)	Changes in cropping practices, rescheduling planting, growing crops requiring less water, and buying crop insurance.	India
Panda (2013)	Water conservation, reducing water use, shift to cotton from rice, change in planting dates, reducing the area of cropping, diversifying crop varieties, diversifying income, and early maturity variety seeds.	India
Khatri-Chhetri et al. (2017)	Rainwater harvesting, laser land levelling, furrow irrigated bed planting, drip irrigation, cover crops method, zero tillage/minimum tillage, site-specific integrated nutrient management, green manuring, leaf colour chart, intercropping with legumes, agroforestry, concentrate feeding for livestock, fodder management, integrated pest management, and weather-based crop agro-advisory.	India
Banerjee (2014)	Crop diversification, rainwater harvesting, check dams, farm pond, and natural resource management.	India
Kattumuri et al. (2017)	Irrigation provisioning, a shift in cropping patterns, mixed-cropping, agroforestry, diversified livestock holdings, leaving croplands fallow, selling assets such as livestock and trees, and migration.	India
Tripathi and Mishra (2017)	Short-duration varieties, change in sowing and harvesting timing, inter-cropping, change in the cropping pattern, investment in irrigation, and agroforestry.	India
Tripathi and Mishra (2017)	Crop diversification, agroforestry, and increased groundwater for irrigation.	India
Aggarwal et al. (2019)	Dug well, tube well, rainwater harvesting, drip irrigation, sprinkler irrigation, farmyard manure, vermicompost, residue incorporation, broad bed furrow, minimum tillage, use of improved seeds, crop diversification, green manuring, gully control structure, legume integration and mulching.	India
Swami and Parthasarathy (2020)	Change in planting/harvesting date (PHD), crop diversification, short-duration crops (SDC), drought-resistant crops (DRC), and micro irrigation.	India
Singh (2020)	Drought tolerant crop varieties, early maturing seed varieties.	India

the determinants of CSA adoption, they were classified into six broad categories: (1) access to extension services; (2) socioeconomic characteristics; (3) experience or perception of climate change and shocks; (4) attitude and behaviour towards risk; (5) farm characteristics; and (6) other factors.

3.5 Access to Extension Services

Access to extension services means farmers can get help from the government, non-government organizations, and other groups to undertake adoption in their farming activities. Access to extension services is the backbone of farmer adoption. Institutional factors cover the extension support on crop and livestock production, access to information on climate change, and access to credit enhancement (Deressa et al., 2009; Alauddin and Sarker, 2014). Access to extension support from government sources empowers farmers with knowledge and awareness, and farmers getting extension support to adopt more proactively than those who do not (Abid et al., 2015; Sardar et al., 2021). The agricultural extension helps farmers by providing them access to training and workshops and spreading the word about how good the CSA measures are. Institutions are very important for giving farmers the information and knowledge they need to take care of the soil, deal with low fertility and dryland soils, and make the best use of rainwater storage to deal with the lack of water in the region. Farmers with solid linkages, regular interaction with extension agents, and easy access to extension services in the study area could help adopt CSA practices. It was found that poor farmers depended a lot on government help, extension services, and information about climate change to decide how to farm. Households with access to food aid, farm support, a radio, a toilet, and electricity, as well as those with fertile soil, modern tools and equipment, informal sources of credit, and information about climate change, were better able to adopt climate change (Bryan et al., 2009). Hassan and Nhemachena (2008) identified a range of extension factors that influence farmers' decisions; the significant factors that determine the adoption decision are access to resources, institutional factors, technology and farm assets (labour, land, and capital), access to markets, and extension and credit services, which critically persuade farmers to adopt climate change activities. Marenya and Barrett (2007) found that factors such as access to credit, generation of off-farm income, and membership in social groups positively impacted the adoption of improved Natural resource management (NRM) practices. Deressa et al. (2009) found that farmers were more likely to adapt well to climate change if they had access to credit and extension services, had larger farms, had more education, and had more social capital.

Government and non-government organizations help farmers find ways to make money outside of farming so they can have more than one source of income and keep farming even when the weather is unpredictable. Off-farm livelihood generation could help farmers to adopt climate-resilient agricultural practices (Gbetibouo, 2009). Community assets, such as access to government technical services during droughts, the concentration of continuous residential areas, and the number of lateral canals, also affect adoption in drought-prone areas (Alauddin and Sarker, 2014; Yang-jie et al., 2014).

According to Muller and Shackleton (2014), a significant obstacle for farmers in adapting to climate change is the need for more access to information about climate change and the various options available for adoption. Access to market information, weather forecasting information, and agricultural activity information positively impacts CSA adoption (Ritho et al., 2012; Elum et al., 2017). The likelihood of a change in the crop calendar as a means of adoption increases when farmers have access to information about temperature and rainfall (Deressa et al., 2009). Informal, formal, endogenous, and externally initiated institutions are interdependent and equally important in the processes of adoption and innovation. Connectivity and interaction between these institutions can benefit the adoption process (Rodima-Taylor et al., 2012; Abid et al., 2015).

Through communities of practice, formal institutions play a crucial role in building place-based capacity for strategies to deal with climate change and adapt to it in agriculture. These institutions help farmers learn how to change on their own. Adoption practices can be improved if there are good links between formal and informal institutions that work on climate change and if they share information. Public and private institutions play a big part in helping people to adopt CSA. Better connections between government agencies and beneficiaries can improve how policies are made and carried out, making it easier for smallholders to adopt. If adoption policies are made by the central government alone, without the help of local governments, it could mean that farmers do not get the most out of them (Islam and Nursey-Bray, 2017; Mubaya and Mafongoya, 2017; Ampaire et al., 2017; Khanal et al., 2019).

Ojo and Baiyegunhi (2020) expounded that credit availability is a significant factor in adoption. The source of credit and the distance from the source of credit are significant determinants of adoption. Tessema et al. (2013) say that making people in rural areas more aware of credit options can help them adapt to climate change. Makate et al. (2019) say that it takes money and institutional support to implement adoption measures that can lessen the bad effects of climate change. Other studies assert that access to credit from formal and non-formal institutions could enhance the adoption of CSA practices (Tessema et al., 2013; Balew et al., 2014; Adenle et al., 2015; Swami and Parthasarathy, 2020).

Adenle et al. (2015) found that agricultural innovation systems will be one of the most important ways for developing countries to fight climate change. Policy and institutional change are vital. Many developing countries spend only a small amount of their money on research and development of technologies that could help them deal with the effects of climate change on agricultural production and adapt to them. Weak infrastructure, limited research capacity, a lack of credit facilities, and an inability to share technology make it harder to use innovation to solve the problems caused by climate change.

Mubaya and Mafongoya (2017) used a qualitative method to learn about institutional and structural issues and the role of institutions and institutional arrangements in helping people adapt locally. In the local-level adoption strategy, they have found three ways to change: both public and private institutions in the study areas play a big part in making adoption easier; there is a clear difference between the functions of public, private, and civic institutions; institutions and informal arrangements between institutions help people to adopt together.

Jain et al. (2015) found that infrastructural development, such as access to irrigation and willingness to take risks, enhances the adoption decision. Farmers who had more resources and access to irrigation used more irrigation-related adoption strategies. Farmers who were poorer and had less reliable access to irrigation were more likely to change the planting date as the farming strategy during water scarcity.

Distance between extension offices and markets significantly impacts the adoption of CSA practices. The closer the distance, the higher the adoption rate. Mazhar et al. (2021) and Aryal et al. (2018) found a significant negative association between the distance from the market and the adoption of CSA practices. When the distance to the market is shorter, it is probable that farmers will implement various agricultural practices such as crop diversification, stress-tolerant varieties, laser land levelling, minimum tillage, and site-specific nutrient management. According to Jena et al.'s (2023) research, the proximity to the primary market, village market, and agricultural extension office had a positive impact on the adoption of minimum tillage.

3.6   Socioeconomic Characteristics

The socioeconomic characteristics include the following subgroups: household demographic characteristics such as age, gender, household size, farming experience, caste, education level of the household, and family members. The economic characteristics include the asset accumulation and income of the farmer. The secondary occupation and income of the household also come under the category of the economic status of the farmers. Migration and the remittances received from it fall under the economic livelihood activities of the farmer. A series of studies advocate the significant impact of socioeconomic factors on farmers' adoption decisions for climate-smart agriculture.

The age of the household and farming experience have contrasting effects on the adaptive decision. The literature argues that the older the farmer, the more experience he has. The longer the farming experience, the more likely farmers are to take adaptive measures because farmers with vast experience could have extensive observation-based knowledge of climate change and adoption (Deressa et al., 2009). A higher age group of household heads had adopted agroforestry and irrigation-related adoption practices. The study conducted by Khatri-Chhetri et al. (2017) showed a statistically significant and positive correlation between the age of the household head and the implementation of integrated nutrient management, pest management, laser land levelling, and crop insurance practices. Hassan and Nhemachena (2008) found that experienced farmers are more likely to adapt to climate change. Another group of studies argues that the older the heads of household, the more conservative they will be. So, there is a negative relationship between age and the adoption of climate change (Shiferaw and Holden, 1998). Farmers who have been farming for a long time have gained local knowledge that helps them deal with problems better than younger farmers. This has resulted in their preferring to use traditional knowledge rather than embrace modern techniques (Nyong et al., 2007; Maguza-Tembo et al., 2017).

Gender plays a major role in adopting CSA practices (Asfaw and Admassie, 2004; Ngigi et al., 2017). In a household, the adoption of CSA strategies differs between the female and male heads. Women are more likely to adopt crop-related strategies, while men are more likely to adopt livestock-related strategies. Gender-differentiated group-based adoption interventions are advocated. Male groups comprise a mix of genders, while female groups comprise only female members. This difference in composition affects perception and belief systems. Implementing gender-based interventions in groups can help both genders adapt more efficiently (Ngigi et al., 2017). Male farmers are more prone to taking risks, adopting new technologies and adapting their farming methods than their female counterparts (Asfaw and Admassie, 2004). Meher et al. (2016) say that women farmers cannot use better farming methods because they do not have enough access to information, land, and other resources. However, in some regions, an opposing perspective suggests that women are more likely to take on adaptive measures because they are actively and intensively involved in farming practices (Nhemachena and Hassan, 2007; Jost et al., 2016).

Adger et al. (2003) explored the issue of climate change adoption in developing countries. The authors pointed out that as compared to developed countries, developing countries have lesser money and weaker infrastructure, making it harder for the latter to adopt new ideas. Smit and Pilifosova (2003) explored the relationship between climate change adoption, sustainable development, and equity. They argue that socioeconomic characteristics play an important role in enhancing the community's and household's adaptive capacity. The weak adaptive capacity is due to a lack of proper economic resources. Uddin et al. (2014) and Abid et al. (2015) found that primary and off-farm income are crucial factors affecting farmers' adoption of climate change. Burnham and Ma (2017) say that a farmer's ability to adapt can be boosted by having a higher income and the fact of having done it before.

Broadly, the studies of Deressa et al. (2009), Gbetibouo (2009), Mertz et al. (2009), Abid et al. (2015), Bahinipati (2015), Elum et al. (2017), and Swami and Parthasarathy (2020) reported that household factors such as the level of education of the household head, household size, land size, ownership of tube wells, the gender of the head of the household, the age of the farmer, and the property of the household are the major factors that affect how well they can adapt to climate change. Marenya and Barrett (2007) found that household characteristics such as age, gender, education level, and farm size had a mixed or negative impact on adapting to climate change.

Education promotes climate-smart agriculture. Educated farmers will learn about new technologies and share them to practice resilience to climate change. Farmer's view of climate change improves with education. This shows that educated farmers comprehend climate change better and use innovative communication methods to acquire knowledge (Abid et al., 2019). Deressa et al. (2009; 2011) found that educated farmers had adopted soil conservation and changing planting dates. Meher et al. (2016) found that education levels positively impact the adoption of crop rotation. Other studies support the positive effect of adoption on CSA practices (Maddison, 2007; Kumar, 2011; Kakumanu et al., 2016).

The social category is one of the covariates in adopting climate change. In Indian society, the social structure plays a critical role, especially in accessing information and facilities. Caste is one of the social capitals that affect the inequality of public spheres in rural communities of Southern Asia (Aryal and Holden, 2012; Holden et al., 2013; Aryal et al., 2018). There are four categories of caste present in society. Among them, those who belong to the bottom of the hierarchy are not able to access information and opportunities (Birthal et al., 2015). The lower rung in the hierarchy of the caste system, especially the scheduled tribes (ST) and scheduled castes (SC), is less adaptive than the higher castes (Yamano et al., 2015). Farmers from the general caste group are likely to adopt more than others. Khatri-Chettri (2017) found that the general caste farmers were not interested in adopting weather-based crop agro-advisories and crop insurance. The probability of the general caste group adopting CSAPs is higher than the likelihood of the backwards and scheduled caste groups adopting them. Farmers who are categorized under the general caste group have a higher tendency to adopt stress-tolerant seeds and site-specific nutrient management. However, they are less inclined to adopt minimum tillage (Aryal et al., 2017). A greater proportion of households not belonging to the SC/ST category identified early maturity and disease resistance as significant characteristics. SC/ST households gave a high rank to lodging tolerance, whereas resistance to pests received a higher ranking from households belonging to other castes (Krishna and Veettil, 2022).

Inter-state and intra-state migration are important in determining the adoption of CSA practices. The head of the household or other family members relocates to seek better income opportunities. The money they send back home as remittances helps improve the household's ability to adjust to changing circumstances (Bahinipati et al., 2021). Migrating households have a comparative advantage over non-migrating households regarding adaptive capacity (Jha et al., 2018).

3.7   Experience/Perception of Climate Change

Scholars have argued that the adoption to climate change in third-world countries would largely depend on the experiences that local communities have had in dealing with climate-related risks in the past (Adger et al., 2003). The increase in temperature and decrease in precipitation affect the adoption behaviour of a farmer (Deressa et al., 2011). There are studies shows the perception of climate change positively and significantly impacted adopting CSA practices (Praveen and Ramachandran, 2015).

Gbetibouo (2009) found that farmers in Limpopo Basin are aware of the changes in temperature, rainfall, and weather patterns and have identified climate change as the cause of these changes. Accordingly, farmers modified their planted crops, increase planation of green plants that survive in dry conditions, adjusted water use, and protected the soil.

Jain et al. (2015) stated that farmers' adoption choices depended on their perception of the weather and willingness to take risks. Farmers who perceive that temperatures are increasing and rainfall is decreasing tend to adopt adoption practices more positively than others (Gandure et al., 2013; Tambo and Abdoulaye, 2013). Farmers who experience increased pests and diseases are more likely to adopt adoption measures (Banerjee, 2014).

Bryan et al. (2009) found that the perception of an increase in temperature and rainfall had a significant positive effect on adopting measures to deal with climate change. Mertz et al. (2009) have studied and found that people in the rural Sahel know about climate change and are trying to find ways to adapt. Abid et al. (2015) evaluated that most farm households are aware of climate change and are changing their farming methods. Due to climate change, 58% of farm households have changed the way of farming according to their climatic condition.

Adger et al. (2003) pointed out that compared to developed countries, developing countries face multiple challenges in adopting, due to their limited resources and weaker infrastructure. The climate change vulnerable sections, such as farmers, fishermen, coastal residents, and urban dwellers, would be self-directed and aided by their social connections and available resources to adapt to climate change in agriculture. Farmers who had experienced a drought in the last five years positively impacted the adoption decision (Bryan et al., 2009).

Farmers' perceptions of rising temperatures and decreasing rainfall amounts were consistent with meteorological data. The empirical result also shows farmers adopted crop and farm diversification, agroforestry, mixed farming, intensified irrigation, planting rotation, soil moisture conservation, and cultivating short-season varieties and drought-tolerant crops with the change in climate. Climate change is due to human and natural activities. A farmer cannot take an adoption action if farmers fail to grasp the idea of anthropogenic climate change and its detrimental effects on agriculture. Farmers have witnessed weather events such as droughts and floods more reliably than others (Li et al., 2017). Perceived risk is the core concept of reacting to climate change. Risk perception leads to the belief in "adverse effects for valued objects". The relationship between perceived risk and public response to climate change has been consistent regarding importance and predictive strength (Arbuckle et al., 2015).

Traditional knowledge among the farmers influences the perception of climate change (Boillat and Berkes, 2013). The perception of climate change and the scientific evidence on climate change have been consistent. Indigenous patterns of understanding climate change phenomena appear constant among indigenous farmers. They have used their indigenous awareness to embrace climate change (Mekonnen et al., 2018).

3.8   Attitude and Behaviour towards Risk

Grothmann and Patt (2005) examined the role of human cognition in individual adoption to climate change. The individual experience is drawn from their perception of risk, social norms, and access to and availability of information and resources that influence the adoption of climate change.

Jim et al. (2012) found that capacity-enhancement activities and profit-oriented behaviour positively impacted farmers' adoption of certified seed technology and integrated crop management practices. Truelove et al. (2015) found that drought risk perceptions, efficacy beliefs, village identification, perceived descriptive norms, social networks, cultural values, and psychological factors such as risk perception and self-efficacy had a major role to play in the adoption of sustainable agriculture. Farmers who thought the risk of the drought was high were not expected to do anything to protect themselves unless they felt like they could handle the threat. Farmers who thought the risk of the drought was low were not expected to do anything to protect themselves, no matter how good they were.

Li et al. (2017) found that the main drivers of farmers' adoption behaviour are financial motives and managerial considerations. Farmers who aim to increase profits and sales, acquire farm ownership, manage larger land areas, possess innovative personalities, and have access to information from socio-agricultural networks, and are more likely to adopt adaptive measures. According to Burnham and Ma (2017), having a higher income and previous adoption experiences can increase a farmer's perceived self-efficacy. The availability of farm labour is also a factor that contributes to perceived self-efficacy, an important determinant of adoption.

Azadi et al. (2019) found a complicated link between farmers' adoption behaviours and their overall beliefs about climate change, including how they see risks, how far away they are from the problem, how much they trust it, and how important the risk is to them. Climate risk perception, trust, and psychological distance drove farmers' adoption behaviours more effectively. Farmers had a sound awareness of climate change activities and associated risks, and this understanding of risk led to the trial of the adoption option. When farmers experience climate change and perceive it as a "problem", their willingness to take action is activated (Arbuckle et al., 2015).

3.9   Farm Characteristics

The farm plays a major role in facilitating the adoption of CSA practices. The farm characteristics that most studies have mentioned are the location of the land, size of the land, rainfed/irrigated land, soil quality, soil characteristics, availability of irrigation, availability of input, and availability of farm power. Farmers with a large piece of land are likely to adopt CSA. The large land holdings trigger farmers to invest in improved technology (Rajendran, 2020). A few studies reported mixed results about the connection between land holding size and adoption of CSA practices. The studies (Kafle, 2011; Pongvinyoo et al., 2014; Xie et al., 2016; Kunzekweguta et al., 2017) signify the positive correlation between the size of the land holdings and the adoption of CSA practices. However, other studies, Okon and Idiong (2016) and Digal and Placencia (2019), establish a negative relation between the land holding size and the adoption of CSA practices. Alauddin and Sarker (2014) and Yang-jie et al. (2014) reported that farm infrastructure plays a key role in determining farmers' adoption strategies. Access to electricity on the farm can help a farmer adopt water-saving strategies. In contrast, the lack of electricity may result in rescheduled planting and water-saving adoptions. Community assets, such as access to government technical services during droughts, the concentration of continuous residential areas, and the number of lateral canals, also affect adoption in drought-prone areas.

References

Abdur Rashid Sarker, M., Alam, K., and Gow, J. (2013) 'Assessing the determinants of rice farmers' adaptation strategies to climate change in Bangladesh', *International Journal of Climate Change Strategies and Management*, 5(4), pp. 382–403. https://doi.org/10.1108/IJCCSM-06-2012-0033

Abeysingha, N. S., Singh, M., Islam, A. and Sehgal, V. K. (2016) 'Climate change impacts on irrigated rice and wheat production in Gomti River basin of India: A case study', *Springer Plus*, 5, pp. 1–20.

Abid, M., Scheffran, J., Schneider, U. A. and Ashfaq, M. (2015) 'Farmers' perceptions of and adaptation strategies to climate change and their determinants: The case of Punjab province, Pakistan', *Earth System Dynamics*, 6(1), pp. 225–243.

Abid, M., Scheffran, J., Schneider, U. A., & Elahi, E. (2019) 'Farmer perceptions of climate change, observed trends and adaptation of agriculture in Pakistan', *Environmental Management*, 63, pp. 110–123. https://doi.org/10.1007/s00267-018-1113-7.

Adams, R. M., Rosenzweig, C., Peart, R. M., Ritchie, J. T., McCarl, B. A., Glyer, J. D., Curry, R. B., Jones, J. W., Boote, K. J. and Allen, L. H. (1990) 'Global climate change and US agriculture', *Nature*, 345(6272), pp. 219–224. https://doi.org/10.1038/345219a0.

Adenle, A. A., Azadi, H. and Arbiol, J. (2015) 'Global assessment of technological innovation for climate change adaptation and mitigation in developing world', *Journal of Environmental Management*, 161, pp. 261–275.

Adger, W. N., Arnell, N. W. and Tompkins, E. L. (2005) 'Successful adaptation to climate change across scales', *Global Environmental Change*, 15(2), pp. 77–86.

Adger, W. N., Huq, S., Brown, K., Conway, D. and Hulme, M. (2003) 'Adaptation to climate change in the developing world', *Progress in Development Studies*, 3(3), pp. 179–195.

Alam, G. M. M., Alam, K. and Mushtaq, S. (2017) 'Climate change perceptions and local adaptation strategies of hazard-prone rural households in Bangladesh', *Climate Risk Management*, 17, pp. 52–63. https://doi.org/10.1016/j.crm.2017.06.006

Alauddin, M. and Sarker, M. A. R. (2014) 'Climate change and farm-level adaptation decisions and strategies in drought-prone and groundwater-depleted areas of Bangladesh: An empirical investigation', *Ecological Economics*, 106, pp. 204–213.

Ampaire, E. L., Jassogne, L., Providence, H., Acosta, M., Twyman, J., Winowiecki, L. and Van Asten, P. (2017) 'Institutional challenges to climate change adaptation: A case study on policy action gaps in Uganda,' *Environmental Science & Policy*, 75, pp. 81–90.

Antle, J. M. (1995) 'Climate change and agriculture in developing countries', *American Journal of Agricultural Economics*, 77(3), pp. 741–746.

Anwar, M. R., Liu, D. L., Macadam, I. and Kelly, G. (2013) 'Adapting agriculture to climate change: A review', *Theoretical and Applied Climatology*, 113(1–2), pp. 225–245.

Arbuckle Jr, J. G., Morton, L. W. and Hobbs, J. (2015) 'Understanding farmer perspectives on climate change adaptation and mitigation: The roles of trust in sources of climate information, climate change beliefs, and perceived risk', *Environment and Behavior*, 47(2), pp. 205–234.

Aryal, J. P. and Holden, S. T. (2012) 'Livestock and land share contracts in a Hindu society', *Agricultural Economics*, 43(5), pp. 593–606.

Aryal, K., Poudel, S., Chaudhary, R. P., Chettri, N., Ning, W., Shaoliang, Y. and Kotru, R. (2017) 'Conservation and management practices of traditional crop genetic diversity by the farmers: A case from Kailash Sacred Landscape, Nepal,' *Journal of Agriculture and Environment*, 18, pp. 15–28.

Aryal, J. P., Jat, M. L., Sapkota, T. B., Khatri-Chhetri, A., Kassie, M., Rahut, D. B. and Maharjan, S. (2018) 'Adoption of multiple climate-smart agricultural practices in the gangetic plains of Bihar, India', *International Journal of Climate Change Strategies and Management*, 10(3), pp. 407–427.

Asfaw, A. and Admassie, A. (2004) 'The role of education on the adoption of chemical fertiliser under different socioeconomic environments in Ethiopia', *Agricultural Economics* 30(3), pp. 215–228.

Azadi, H., Moghaddam, S. M., Burkart, S., Mahmoudi, H., Van Passel, S., Kurban, A. and Lopez-Carr, D. (2021) 'Rethinking resilient agriculture: From climate-smart agriculture to vulnerable-smart agriculture', *Journal of Cleaner Production*, 319, p. 128602.

Azadi, Y., Yazdanpanah, M. and Mahmoudi, H. (2019) 'Understanding smallholder farmers' adaptation behaviors through climate change beliefs, risk perception, trust, and psychological distance: Evidence from wheat growers in Iran', *Journal of Environmental Management*, 250, p. 109456.

Bahinipati, C. S., Kumar, V. and Viswanathan, P. K. (2021) 'An evidence-based systematic review on farmers' adaptation strategies in India', *Food Security*, 13(2), pp. 399–418.

Bahinipati, C. S. and Venkatachalam, L. (2015) 'What drives farmers to adopt farm-level adaptation practices to climate extremes: Empirical evidence from Odisha, India', *International Journal of Disaster Risk Reduction*, 14, pp. 347–356.

Balew, S., Agwata, J. and Anyango, S. (2014) 'Determinants of adoption choices of climate change adaptation strategies in crop production by small scale farmers in some regions of central Ethiopia,' *Journal of Natural Sciences Research*, 4(4), pp. 78–93, https://www.iiste.org/Journals/index.php/JNSR/article/view/11183/11484.

Bandara, J. S. and Cai, Y. (2014) 'The impact of climate change on food crop productivity, food prices and food security in South Asia', *Economic Analysis and Policy*, 44(4), pp. 451–465.

Banerjee, R. R. (2014) 'Farmers' perception of climate change, impact and adaptation strategies: A case study of four Villages in the semi-arid regions of India', *Natural Hazards*, 75(3), pp. 2829–2845.

Banerjee, G. D. and Banerjee, S. (2014) 'Crop diversification: An exploratory analysis', In *Diversification of agriculture in eastern India*, New Delhi: Springer India, pp. 37–57

Belay, A., Recha, J. W., Woldeamanuel, T. and Morton, J. F. (2017) 'Smallholder farmers' adaptation to climate change and determinants of their adaptation decisions in the Central Rift Valley of Ethiopia,' *Agriculture & Food Security*, 6, pp. 1–13.

Birthal, P. S., Kumar, S., Negi, D. S. and Roy, D. (2015) 'The impacts of information on returns from farming: Evidence from a nationally representative farm survey in India', *Agricultural Economics*, 46(4), pp. 549–561.

Boillat, S. and Berkes, F. (2013) 'Perception and interpretation of climate change among quechua farmers of Bolivia', *Ecology and Society*, 18(4), p. 21.

Bryan, E., Deressa, T. T., Gbetibouo, G. A. and Ringler, C. (2009) 'Adaptation to climate change in Ethiopia and South Africa: Options and constraints', *Environmental Science and Policy*, 12(4), pp. 413–426.

Burnham, M. and Ma, Z. (2017) 'Climate change adaptation: Factors influencing Chinese smallholder farmers perceived self-efficacy and adaptation intent', *Regional Environmental Change*, 17, pp. 171–186.

Campbell, B. M., Thornton, P., Zougmoré, R., Van Asten, P. and Lipper, L. (2014) 'Sustainable intensification: What is its role in climate smart agriculture?'. *Current Opinion in Environ Sustain*, 8, pp. 39–43.

Deressa, T. T., Hassan, R. M. and Ringler, C. 2011 'Perception of an adaptation to climate change by farmers in the Nile basin of Ethiopia,' *The Journal of Agricultural Science*, 149(1), pp. 23–31. https://doi.org/10.1017/S0021859610000687.

Deressa, T. T., Hassan, R. M., Ringler, C., Alemu, T. and Yesuf, M. (2009) 'Determinants of farmers' choice of adaptation methods to climate change in the Nile basin of Ethiopia', *Global Environmental Change*, 19(2), pp. pp. 248–255.

Digal, L. N. and Placencia, S. G. P. (2019) 'Factors affecting The adoption of organic rice farming: The case of farmers in M'lang, North cotabato, Philippines', *Organic Agriculture*, 9, pp. 199–210.

Dinesh, D., Frid-Nielsen, S., Norman, J., Mutamba, M., Loboguerrero Rodriguez, A. M. and Campbell, B. M. (2015) 'Is Climate-Smart Agriculture effective? A review of selected cases', *CCAFS Working Paper*. https://cgspace.cgiar.org/items/c18981fa-1cae-4005-9d8c-a4638ffc8ed7.

Elum, Z. A., Modise, D. M. and Marr, A. (2017) 'Farmer's perception of climate change and responsive strategies in three selected provinces of South Africa', *Climate Risk Management*, 16, pp. 246–257. https://doi.org/10.1016/j.crm.2016.11.001

Fankhauser, S., Smith, J. B. and Tol, R. S. (1999) 'Weathering climate change: Some simple rules to guide adaptation decisions', *Ecological Economics*, 30(1), pp. 67–78.

FAO (2010) 'Climate-smart agriculture: Policies, practices and financing for food security, adaptation and mitigation', *FAO*, 49.

FAO. (2012) 'Mainstreaming climate-smart agriculture into a broader landscape approach. Background Paper for the Second Global Conference on Agriculture', *Food Security and Climate Change Hanoi*, Vietnam, 3-7 September 2012.

FAO. (2013) Climatesmart agriculture. https://www.fao.org/4/i3325e/i3325e.pdf

FAO. (2015) 'Climate change and food systems: global assessments and implications for food security and trade'. Rome.

FAO. (2017) 'FAO Source Book'. Retrieved from https://www.fao.org/climate-smart-agriculture-sourcebook/en.

Feder, G., Just, R. E. and Zilberman, D. (1985) 'Adoption of agricultural innovations in developing countries: A survey', *Economic Development and Cultural Change*, 33(2), pp. 255–295.

Fosu-Mensah, B. Y., Vlek, P. L. G. and MacCarthy, D. S. (2012) 'Farmers' perception and adaptation to climate change: A case study of Sekyedumase district in Ghana', *Environment, Development and Sustainability*, 14(4),pp. 495–505. https://doi.org/10.1007/s10668-012-9339-7

Funk, C., Raghavan Sathyan, A., Winker, P. and Breuer, L. (2020) 'Changing climate - changing livelihood: Smallholder's perceptions and adaption strategies', *Journal of Environmental Management*, 259(October), p. 109702. https://doi.org/10.1016/j.jenvman.2019.109702

Gandure, S., Walker, S. and Botha, J. J. (2013) 'Farmers' perceptions of adaptation to climate change and water stress in a South African rural community', *Environmental Development*, 5(1), pp. 39–53.

Gbetibouo, G. A. (2009) 'Understanding farmers' perceptions and adaptations to climate change and variability: The case of the Limpopo Basin, South Africa (Vol. 849).' *International Food Policy Research Institute*.

Ghosh, M. (2019) 'Climate-smart agriculture, productivity and food security in India,' *Journal of Development Policy and Practice*, 4(2), pp. 166–187.

Grothmann, T. and Patt, A. (2005) 'Adaptive capacity and human cognition: The process of individual adaptation to climate change', *Global Environmental Change*, 15(3), pp. 199–213.

Guiteras, R. (2009) *The impact of climate change on Indian agriculture*. Manuscript, Department of Economics, University of Maryland, College Park, Maryland.

Gupta, R., Somanathan, E. and Dey, S. (2017) 'Global warming and local air pollution have reduced wheat yields in India', *Climatic Change*, 140(3–4), pp. 593–604.

Hassan, R. M. and Nhemachena, C. (2008) 'Determinants of African farmers' strategies for adapting to climate change: Multinomial choice analysis,' *African Journal of Agricultural and Resource Economics*, 2(1), pp. 83–104.

Hitz, S. and Smith, J. (2004) 'Estimating global impacts from climate change,' *Global Environmental Change*, 14(3), pp. 201–218. https://doi.org/10.1016/j.gloenvcha.2004.04.010

Holden, M. P., Duff-Canning, S. J. and Hampson, E. (2015) 'Sex differences in the weighting of metric and categorical information in spatial location memory,' *Psychological Research*, 79, pp. 1–18.

IPCC. (2001) *Climate change 2001: Impacts, adaptation, and vulnerability: contribution of Working Group II to the third assessment report of the Intergovernmental Panel on Climate Change*. Cambridge, UK; New York: Cambridge University Press.

IPCC. (2014) IPCC 2014. Clim. Chang. 2014 Synth. Report. Contrib. Work. Groups I, II III to Fifth Assess. Rep. Intergov. Panel Clim. Chang.

Islam, M. T. and Nursey-Bray, M. (2017) 'Adaptation to climate change in agriculture in Bangladesh: The role of formal institutions', *Journal of Environmental Management*, 200, pp. 347–358.

Jain, M., Naeem, S., Orlove, B., Modi, V. and DeFries, R. S. (2015) 'Understanding the causes and consequences of differential decision-making in adaptation research: Adapting to a delayed monsoon onset in Gujarat, India', *Global Environmental Change*, 31, pp. 98–109.

Jena, P. R., Tanti, P. C. and Maharjan, K. L. (2023) 'Determinants of adoption of climate resilient practices and their impact on yield and household income,' *Journal of Agriculture and Food Research*, 14(June). https://doi.org/10.1016/j.jafr.2023.100659

Jha, C. K., Gupta, V., Chattopadhyay, U. and Amarayil Sreeraman, B. (2018) 'Migration as adaptation strategy to cope with climate change: A study of farmers' migration in rural India', *International Journal of Climate Change Strategies and Management*, 10(1), pp. 121–141.

Jim, M., Villano, R. and Fleming, E. (2012) 'Factors influencing farmers' adoption of modern rice technologies and good management practices in the Philippines', *Agricultural Systems*, 110, pp. 41–53.

Jost, C., Kyazze, F., Naab, J., Neelormi, S., Kinyangi, J., Zougmore, R. and Kristjanson, P. (2016) 'Understanding gender dimensions of agriculture and climate change in smallholder farming communities', *Climate and Development*, 8(2), pp. 133–144.

Kafle, B. (2011) 'Factors affecting adoption of organic vegetable farming in Chitwan district, Nepal', *World Journal of Agricultural Sciences*, 7(5), pp. 604–606.

Kaiser, H. M., Riha, S. J., Wilks, D. S., Rossiter, D. G. and Sampath, R. (1993) 'A farm-level analysis of economic and agronomic impacts of gradual climate warming,' *American Journal of Agricultural Economics*, 75(2), pp. 387–398. https://doi.org/10.2307/1242923

Kakumanu, K. R., Kuppanan, P., Ranganathan, C. R., Shalander, K. and Amare, H. (2016) 'Assessment of risk premium in farm technology adoption as a climate change adaptation strategy in the dryland systems of India', *International Journal of Climate Change Strategies and Management*.

Katz Elihu. (1961) 'The social itinerary of technical change: Two studies on the diffusion of innovation', *Human Organization*, 20(2), pp. 70–82.

Khanal, U., Wilson, C., Lee, B. L., Hoang, V. N. and Managi, S. (2019) 'Influence of payment modes on farmers' contribution to climate change adaptation: Understanding differences using a choice experiment in Nepal', *Sustainability Science*, 14(4), pp. 1027–1040.

Khatri-Chhetri, A. (2017) Progressing towards climate resilient agriculture: Top ten success stories from CCAFS in South Asia.

Khatri-Chhetri, A., Aggarwal, P. K., Joshi, P. K. and Vyas, S. (2017) 'Farmers' prioritization of climate-smart agriculture (CSA) technologies,' *Agricultural Systems*, 151, pp. 184–191.

Knowler, D. and Bradshaw, B. (2007) 'Farmers' adoption of conservation agriculture: A review and synthesis of recent research', *Food Policy*, 32(1), pp. 25–48.

Krishna, V. V. and Veettil, P. C. (2022) 'Gender, caste, and heterogeneous farmer preferences for wheat varietal traits in rural India', *PLoS ONE*, 17(8 August), pp. 1–24. https://doi.org/10.1371/journal.pone.0272126

Kumar, K. S. K. (2011) 'Climate sensitivity of Indian agriculture: Do spatial effects matter?', *Cambridge Journal of Regions, Economy and Society*, 4(2), pp. 221–235.

Kumar, K. K. and Parikh, J. (2001) 'Indian Agriculture and climate sensitivity', *Global Environmental Change*, 11(2), pp. 147–154.

Kumar, Y., Singh, R., Kumar, A., Singh, V. and Galav, A. (2018) 'Climate smart agriculture: Approach and perspectives for the rice-wheat crop rotation system in IGP (Indo-gangetic plains)', *Current Journal of Applied Science and Technology*, 28(4), pp. 1–6. https://doi.org/10.9734/cjast/2018/37451

Kunzekweguta, M., Rich, K. M. and Lyne, M. C. (2017) 'Factors affecting adoption and intensity of conservation agriculture techniques applied by smallholders in Masvingo district, Zimbabwe', *Agrekon*, 56(4), pp. 330–346. https://doi.org/10.1080/03031853.2017.1371616

Li, S., Juhász-Horváth, L., Harrison, P. A., Pintér, L. and Rounsevell, M. D. A. (2017) 'Relating farmer's perceptions of climate change risk to adaptation behaviour in Hungary', *Journal of Environmental Management*, 185, pp. 21–30.

Linn J. (2008) 'Energy prices and the adoption of energy-saving technology', *The Economic Journal*, 118(553), pp. 1986–2012.

Lipper, L., Dutilly-Diane, C. and McCarthy, N. (2010) 'Supplying carbon sequestration from West African rangelands: Opportunities and barriers', *Rangeland Ecology and Management*, 63(1), pp155–166. https://doi.org/10.2111/REM-D-09-00009.1

Lipper, L., Thornton, P., Campbell, B. M., Baedeker, T., Braimoh, A., Bwalya, M. and Torquebiau, E. F. (2014) 'Climate-smart agriculture for food security', *Nature Climate Change*, 4(12), pp. 1068–1072.

Lobell, D. B., Sibley, A. and Ivan Ortiz-Monasterio, J. (2012) 'Extreme heat effects on wheat senescence in India', *Nature Climate Change*, 2(3), pp. 186–189. https://doi.org/10.1038/nclimate1356

Maddison, D. (2007) *The perception of and adaptation to climate change in Africa* (Vol. 4308). World Bank Publications. https://openknowledge.worldbank.org/server/api/core/bitstreams/0fb4edaf-6ad0-5488-ae85-a7396ed3bbb1/content.

Maguza-Tembo, F., Mangison, J., Edris, A. K. and Kenamu, E. (2017) Determinants of adoption of multiple climate change adaptation strategies in Southern Malawi: An ordered probit analysis. *Journal of Development and Agricultural Economics*, 9(1), pp. 1–7.

Makate, C., Makate, M. and Mango, N. (2019) 'Wealth-related inequalities in adoption of drought-tolerant maize and conservation agriculture in Zimbabwe', *Food Security*, 11(4), pp. 881–896.

Marenya, P. P. and Barrett, C. B. (2007) 'Household-level determinants of adoption of improved natural resources management practices among smallholder farmers in western Kenya,' *Food Policy*, 32(4), pp. 515–536.

Mazhar, R., Ghafoor, A., Xuehao, B. and Wei, Z. (2021) 'Fostering sustainable agriculture: Do institutional factors impact the adoption of multiple climate-smart agricultural practices among new entry organic farmers in Pakistan?' *Journal of Cleaner Production*, 283, p. 124620.

Meher, M., Mittal, S. and Prasad, N. (2016) 'Farmers coping strategies for climate shock: Is it differentiated by gender?' *Journal of Rural Studies*, 44, pp. 123–131.

Mekonnen, Z., Kassa, H., Woldeamanuel, T. and Asfaw, Z. (2018) 'Analysis of observed and perceived climate change and variability in Arsi Negele district, Ethiopia', *Environment, Development and Sustainability*, 20(3), pp. 1191–1212.

Mendelsohn, R. (2008) 'The impact of climate change on agriculture in developing countries', *Journal of Natural Resources Policy Research*, 1(1), pp. 5–19.

Mendelsohn, R., and Tiwari, D. (2000) *Two essays on climate change and agriculture: A developing country perspective* (No. 145).

Mertz, O., Mbow, C., Reenberg, A. and Diouf, A. (2009) 'Farmers' perceptions of climate change and agricultural adaptation strategies in rural Sahel', *Environmental Management*, 43(5), pp. 804–816.

Mirza, N., Pervez, A., Mahmood, Q., Shah, M. M. and Shafqat, M. N. (2011) 'Ecological restoration of arsenic contaminated soil by Arundo donax L.,' *Ecological Engineering*, 37(12), pp. 1949–1956. https://doi.org/10.1016/j.ecoleng.2011.07.006.

Mishra, D., Sahu, N. C. and Sahoo, D. (2016) 'Impact of climate change on agricultural production of Odisha (India): A Ricardian analysis', *Regional Environmental Change*, 16(2), pp. 575–584.

Mitchell, T., Ibrahim, M., Harris, K., Hedger, M., Polack, E., Ahmed, A. K. and Sajjad Mohammad, S. (2010) 'Climate smart disaster risk management', https://opendocs.ids.ac.uk.

Mubaya, C. P. and Mafongoya, P. (2017) 'The role of institutions in managing local level climate change adaptation in semi-arid Zimbabwe', *Climate Risk Management*, 16, pp. 93–105.

Muller, C. and Shackleton, S. E. (2014) 'Perceptions of climate change and barriers to adaptation amongst commonage and commercial livestock farmers in the semi-arid Eastern cape karoo', *African Journal of Range and Forage Science*, 31(1), pp. 1–12.

Nelson, E., Mendoza, G., Regetz, J., Polasky, S., Tallis, H., Cameron, D. R., Chan, K. M. A., Daily, G. C., Goldstein, J., Kareiva, P. M., Lonsdorf, E., Naidoo, R., Ricketts, T. H. and Shaw, M. R. (2009) 'Modeling multiple ecosystem services, biodiversity conservation, commodity production, and tradeoffs at landscape scales,' *Frontiers in Ecology and the Environment*, 7(1), pp. 4–11. https://doi.org/10.1890/080023

Ngigi, M. W., Mueller, U. and Birner, R. (2017) 'Gender differences in climate change adaptation strategies and participation in group-based approaches: An intra-household analysis from rural Kenya', *Ecological Economics*, 138, pp. 99–108.

Nhemachena, C. and Hassan, R. (2007) 'Micro-level analysis of farmers adaption to climate change in Southern Africa,' *Booksgooglecom*, August, pp. 1–40. http://www.ifpri.org/publication/micro-level-analysis-farmers-adaptation-climate-change-southern-africa-0.

Nordhaus, W. D. (1993) 'Optimal greenhouse-gas reductions and tax policy in the "DICE" model,' *The American Economic Review*, 83(2), pp. 313–317. Papers and Proceedings of the Hundred and Fifth Annual Meeting of the American Economic Association.

Nyong, A., Adesina, F. and Osman Elasha, B. (2007) 'The value of indigenous knowledge in climate change mitigation and adaptation strategies in the African Sahel,' *Mitigation and Adaptation Strategies for Global Change*, 12, pp. 787–797.

Ojo, T. O. and Baiyegunhi, L. J. S. (2020) 'Determinants of credit constraints and its impact on the adoption of climate change adaptation strategies among rice farmers in South-West Nigeria', *Journal of Economic Structures*, 9(1). https://doi.org/10.1186/s40008-020-00204-6.

Okon, U. E. and Idiong, I. C. (2016) 'Factors influencing adoption of organic vegetable farming among farm households in south-south region of Nigeria,' *American-Eurasian Journal of Agricultural and Environmental Sciences*, 16(5), pp. 852–859.

Pongvinyoo, P., Yamao, M. and Hosono, K. (2014) 'Factors affecting the implementation of good agricultural practices (GAP) among coffee farmers in Chumphon province, Thailand', *American Journal of Rural Development*, 2(2), pp. 34–39. https://doi.org/10.12691/ajrd-2-2-3

Praveen, D. and Ramachandran, A. (2015) 'Projected warming and occurrence of meteorological droughts—Insights from the coasts of South India,' *American Journal of Climate Change*, 04(02), pp. 173–179. https://doi.org/10.4236/ajcc.2015.42013.

Rajendran, T. P. (2020) 'Raising farming efficiency for sustained agriculture in Asia and Africa', *Asia-Africa Growth Corridor: Development and Cooperation in Indo-Pacific*, 165–176. https://doi.org/10.1007/978-981-15-5550-3_11. https://link.springer.com/chapter/10.1007/978-981-15-5550-3_11.

Ritho, C., Mbogoh, S. G., Ng'ang'a, S. I., Muiruri, E. J., Nyangweso, P., Kipsat, M. J., Ogada, J. O., Omboto, P. I., Kefa, C., Kubowon, P. C., Cherotwo, F. and Hndambiri, H. K. (2012) 'Assessment of farmers' adaptation to the effects of climate change in Kenya: The case of Kyuso district', *Journal of Economics and Sustainable Development*, 3(12), pp. 52–60.

Rodima-Taylor, D., Olwig, M. F., and Chhetri, N. (2012) 'Adaptation as innovation, innovation as adaptation: An institutional approach to climate change', *Applied Geography*, 33(1), pp. 107–111. https://doi.org/10.1016/j.apgeog.2011.10.011

Rogers, E. M. (1983) Diffusion of Innovations – Chapter 4. In *Diffusion of Innovations* (pp. 160–203). http://ocw.metu.edu.tr/file.php/118/Week9/rogers-doi-ch5.pdf.

Rosenzweig, C. and Hillel, D. (1995) Potential impacts of climate change on agriculture and food supply by,' *Consequences*, 1(2), pp. 23–32. https://doi.org/10.2134/asaspecpub62.c9.

Rosenzweig, C., Iglesias, A., Yang, X. B., Epstein, P. R. and Chivian, E. (2001) 'Climate change and extreme weather events - implications for climate change and extreme weather events - implications for food production, plant diseases, and pests food production, plant diseases, and pests', *Global Change and Human Health*, 2(2), pp. 90–104. http://link.springer.com/10.1023/A:1015086831467

Sardar, A., Kiani, A. K. and Kuslu, Y. (2021) 'Does adoption of climate-smart agriculture (CSA) practices improve farmers' crop income? Assessing the determinants and its impacts in Punjab province, Pakistan', *Environment, Development and Sustainability*, 23, pp. 10119–10140.

Scherr, S. J., Shames, S. and Friedman, R. (2012) 'From climate-smart agriculture to climate-smart landscapes', *Agriculture and Food Security*, 1, pp. 1–15.

Schoengold, K. and Zilberman, D. (2007) 'The economics of water, irrigation, and development', *Handbook of Agricultural Economics*, 3, pp. 2933–2977.

Shiferaw, B. and Holden, S. T. (1998) 'Resource degradation and adoption of land conservation technologies in the Ethiopian highlands: A case study in Andit Tid, North Shewa', *Agricultural Economics*, 18(3), pp. 233–247.

Singh, S. (2020) 'Farmers' perception of climate change and adaptation decisions: A micro-level evidence from Bundelkhand region, India', *Ecological Indicators*, 116, p. 106475. https://doi.org/10.1016/j.ecolind.2020.106475

Smit, B., Burton, I., Klein, R. J. and Street, R. (1999) 'The science of adaptation: a framework for assessment,' *Mitigation and Adaptation Strategies for Global Change*, 4, pp. 199–213.

Smit, B. and Pilifosova, O. (2003) 'Adaptation to climate change in the context of sustainable development and equity', *Sustainable Development*, 8(9), pp. 9.

Swami, D. and Parthasarathy, D. (2020) 'A multidimensional perspective to farmers' decision making determines the adaptation of the farming community', *Journal of Environmental Management*, 264, p. 110487.

Tambo, J. A. and Abdoulaye, T. (2013) 'Smallholder farmers' perceptions of and adaptations to climate change in the Nigerian savanna', *Regional Environmental Change*, 13(2), pp. 375–388.

Tessema, Y. A., Aweke, C. S. and Endris, G. S. (2013) 'Understanding The process of adaptation to climate change by small-holder farmers: The case of East Hararghe zone, Ethiopia', *Agricultural and Food Economics*, 1(1), pp. 1–17.

Tripathi, A. and Mishra, A. K. (2017) 'Knowledge and passive adaptation to climate change: An example from Indian farmers', *Climate Risk Management*, 16(2017), pp. 195–207. https://doi.org/10.1016/j.crm.2016.11.002

Truelove, H. B., Carrico, A. R. and Thabrew, L. (2015) 'A socio-psychological model for analyzing climate change adaptation: A case study of Sri lankan paddy farmers', *Global Environmental Change*, 31, pp. 85–97.

Uddin, M. N., Bokelmann, W. and Entsminger, J. S. (2014) 'Factors affecting farmers' adaptation strategies to environmental degradation and climate change effects: A farm level study in Bangladesh', *Climate*, 2(4), pp. 223–241.

Wheeler, T. and Von Braun, J. (2013) 'Climate change impacts on global food security,' *Science*, 341(6145), pp. 508–513. https://www.science.org/doi/full/10.1126/science.1239402.

Xie, Y., Zhao, H., Pawlak, K. and Gao, Y. (2016) 'The development of organic agriculture in China and The factors affecting organic farming', *Journal of Agribusiness and Rural Development*, 353–361. https://doi.org/10.17306/jard.2015.38

Yamano, T., Rajendran, S. and Malabayabas, M. L. (2015) 'Farmers' self-perception toward agricultural technology adoption: Evidence on adoption of submergence-tolerant rice in Eastern India', *Journal of Social and Economic Development*, 17(2), pp. 260–274. https://doi.org/10.1007/s40847-015-0008-1

Yang-jie, W., Ji-kun, H. and Jin-xia, W. (2014) 'Household and community assets and Farmers' adaptation to extreme weather event: The case of drought in China', *Journal of Integrative Agriculture*, 13(4), pp. 687–697.

Zilberman, D., Zhao, J. and Heiman, A. (2012) 'Adoption versus adaptation, with emphasis on climate change', *Annual Review of Resource Economics*, 4(1), pp. 27–53.

4 Incentives and Barriers in Adoption of Climate-Smart Agriculture

Role of Government and Non-Government Organization to Scale up CSA

4.1 Socio-economic Vulnerability and Climate Smart Agriculture

Natural disasters in India, including cyclones, flash floods, and landslides between 2016 and 2021, caused severe crop damage, impacting more than 36 million hectares of land.[1] This devastation led to financial losses estimated at around $3.75 billion for farmers. As the climate continues to change, it is expected that with a 1.5°C increase in global mean temperature above the pre-industrial level, annual river flooding damages in the country could surge by approximately 49%, while cyclone-related damages might grow by 5.7% (The Hindu, 2022). These climate extremes significantly threaten crop yields, food security, and farm income (Kalli and Jena, 2020, 2022). Therefore, adapting to climate change in agriculture necessitates a comprehensive strategy focused on transitioning to climate-resilient agricultural practices.

Farmers can be enabled to adopt climate-resilient agricultural practices through a combination of external and internal factors. External factors, in particular, play a crucial role in enhancing farmers' adaptive capacities. For example, extension services, including training programmes (Zakaria et al., 2020; Tanti and Jena, 2023), farm field schools (Osumba et al., 2021), and practical demonstrations, significantly increase farmers' awareness and understanding of climate-smart agricultural (CSA) practices (Makate et al., 2019; Mgendi et al., 2022). Access to credit from public and private banks, as well as cooperative societies, further equips farmers to implement these climate-resilient practices (Kangogo et al., 2021).

In addition, subsidies for machinery and seeds are vital in promoting the adoption of CSA practices. These subsidies are particularly helpful for marginal farmers who might otherwise struggle to afford the transition from traditional to improved agricultural methods. The availability of affordable energy sources and the proximity of energy supplies to farm fields are also important. These factors facilitate the use of power-oriented machinery, such as micro-irrigation, sprinkler systems, drip irrigation, and other agricultural equipment (Das et al., 2022). Furthermore, access to clean energy sources enables farmers to harness irrigation resources effectively, supporting sustainable practices

DOI: 10.4324/9781003290445-4

like crop diversification and crop rotation. Various internal factors significantly influence farmers' decisions to adopt CSA practices. These factors include the size of landholding, asset ownership, savings, and income from secondary occupations (Deressa et al., 2011). Economic factors are crucial as they determine the financial capacity of farmers to invest in sustainable agricultural practices, covering the costs of necessary inputs and equipment.

Additionally, household characteristics such as age, gender, farming experience, number of family members, and education level also play an important role in the decision-making process. Educated farmers and those with diverse income sources generally exhibit higher awareness of CSA strategies (Mashi et al., 2022). A deeper understanding of these factors and their interactions can help policymakers and development practitioners more effectively support the adoption of CSA practices.

While theoretically a wide range of factors can influence farmers' decisions about adopting CSA practices, field-based research is essential to identify the specific determinants in particular regions. Considering the unique geographical terrain, soil texture, and local indigenous farming practices of a region, tailored recommendations for CSA practices must be developed. Moreover, the socio-economic composition and cultural beliefs vary significantly across regions. Therefore, there is a need to conduct investigations in climatically vulnerable areas to identify key policy parameters that can be managed to increase the adoption of climate-resilient farming. This forms the primary objective of this study.

Demonstrating the impacts of adopting CSA practices, such as higher farm income or increased crop yield, is crucial for their broader adoption in developing countries. Existing studies have explored these potential impacts in various country contexts. For instance, Mujeyi et al. (2021) examined the socio-economic impact of CSA in Zimbabwe, interviewing 386 households from four districts. Their empirical findings indicated that many smallholder farmers could benefit in terms of food security and revenue from adopting climate-smart agriculture. CSA practices like agroforestry, integrated nutrient management, changes in tillage practices, and crop residue management have been found to mitigate climate change in humid areas. Water management practices, in particular, provide significant food security advantages and mitigation benefits in both dry and humid regions.

Sustainable land management also contributes to food security and environmental benefits through carbon sequestration (Branca et al., 2011). Haq et al. (2021) found a positive correlation between CSA adoption and increased calorie intake in Pakistan's Punjab province, with multiple adoptions positively impacting food diversity and nutritional intake. A recent systematic review by Mizik (2021) explored the potential impact of climate-smart agriculture on marginal farmers, finding that practices like water management and crop rotation have been widely used to increase crop productivity and positively impact both productivity and the environment. Sain et al. (2017) reported that using heat- and water-tolerant maize varieties, pest- and disease-resistant bean varieties, conservation tillage, mulching, agroforestry, crop rotation, contour ditches, stone barriers, surface water reservoirs, and drip irrigation has provided good financial returns for Guatemalan farmers.

However, some studies have found non-significant effects of CSA practices on economic outcomes. Jena (2019) investigated the effect of minimum tillage on maize yield and household income in Kenya, concluding that it did not significantly affect these outcomes but did reduce farm labour use. Jaleta et al. (2016) argued that in Ethiopia, crop management innovations should not be viewed as discrete, stand-alone options; rather, a combination of complementary practices is needed to achieve positive outcomes in crop yield. Thus, the impacts of CSA technology on socio-economic parameters remain inconclusive.

Although previous studies have examined factors leading to CSA adoption and their impacts in various geographical contexts, they have certain limitations. First, very few studies have considered multiple CSA technologies in their analysis. Given the potential complementarities among CSA practices, this study addresses this gap using a multivariate probit model. Second, existing studies often do not explicitly include institutional factors in their adoption models, despite the influential roles of extension training and subsidies for energy and seed. Third, many studies have relied on qualitative or semi-quantitative methods to assess the impact of CSA adoption. Finally, few studies have investigated the adoption and impact of CSA technology in socio-economically and ecologically vulnerable regions that face continuous threats from climate change.

In contrast, the current study focuses on one of the most vulnerable regions of India, where rigorous empirical research on CSA adoption has not yet been conducted. Odisha, an eastern state of India, frequently experiences catastrophic weather events such as floods and droughts, significantly increasing agricultural output unpredictability (Mishra et al., 2016). Table 4.1 lists the series of natural disasters that have plagued the state. Due to regular agricultural shocks resulting in crop loss and unpaid bank credit, some farmers in the state have even committed suicide (Pattanayak and Mallick, 2016; Mohanty and Lenka, 2019). The eastern coastal region of Odisha is prone to sea erosion and soil salinization, while the western part is susceptible to drought and heat waves, leading to crop losses (Panda, 2016; Sahoo and Rath, 2023). Additionally, the socio-economic status of the state is very low, characterized by low per capita consumption expenditures, limited household amenities, high poverty rates, poor health, low female literacy, and poor financial inclusion. Due to socio-economic and climate vulnerability, agricultural practices in the state are less technologically oriented (Mishra et al., 2016).

Apart from researching an underdeveloped, climatically vulnerable, and tribal-dominated landscape, this study contributes to CSA literature in several ways. First, it models multiple CSA practices in a simultaneous estimation procedure to identify the determinants of CSA adoption in Odisha. Second, it explicitly examines the role of institutional and infrastructural factors in scaling up CSA.

Table 4.1 Historical Data on Natural Calamities in Odisha

Years	Natural calamities	Years	Natural calamities
1980	Severe flood, drought	2001	Severe flood, heatwaves
1981	Flood, drought, cyclone, heat waves	2002	Severe drought, heatwaves
1982	Severe flood, drought, a very severe cyclone	2003	Heat waves, flood
1983	Heat waves, flood	2004	Heat waves, flood
1984	Drought, severe floods, cyclones	2005	Severe heatwaves flood, drought
1985	Severe flood, cyclone	2006	Heat waves, severe flood
1986	Drought, severe floods, cyclones	2007	Heatwaves, drought
1987	Drought, severe floods, cyclones	2008	Severe heatwaves flood, drought
1988	Drought, severe heatwaves	2009	Severe drought, heatwaves, cyclones, floods
1989	Drought, cyclones, heat waves, flood	2010	Severe heatwaves, flash floods, and drought
1990	Severe flood	2011	Drought, flood
1991	Severe Flood	2012	Drought, flood
1992	Severe flood, drought	2013	Very severe cyclonic storm `Phailin'/flood
1993	Drought	2014	Flood, very severe cyclonic storm 'Hudhud
1994	Severe flood	2015	Drought, flood & heavy rain
1995	Severe floods, cyclones, heatwaves	2016	Flood and heavy rain, drought
1996	Severe drought, severe heat waves, flood	2017	Flood and heavy rain, drought, pest attack, unseasonal rain
1997	Severe flood and drought	2018	Cyclone "Titli", drought
1998	Severe drought, severe heatwaves, flood	2019	Cyclone "Fani and Bulbul"
1999	Super cyclone, floods, heatwaves	2020	Cyclone "Amphan"
2000	Severe drought, heatwaves	2021	Cyclone "Jawad"

Source: Annual Report on Natural Calamities, Special Relief Commissioner (Government of Odisha), Annual reports of OSDMA (Orissa State Disaster Management Authority), (Das, 2016).

4.2   Materials and Methods

This section outlines the survey design, variable description, and econometric methods employed in the study.

4.2.1   *A Brief Introduction to Odisha*

Odisha, one of the states in India, is located between 17.49'N and 22.34'N latitudes and 81.27'E and 87.29'E longitudes. It features a 480 km long coastline along the Bay of Bengal. Covering an area of 155,707 km², the state stretches 800 km from north to south and 500 km from east to west. Odisha is India's ninth-largest state by land area and the eleventh-largest by population, accounting for 4.7% of the country's land area and 3.7% of its population.

The state is divided into four physiographic zones: the Coastal Plains, the Central Table Land, the Northern Plateau, and the Eastern Ghats. Additionally, Odisha is categorized into ten agro-climatic zones: North Western Plateau, North Central Plateau, North Eastern Coastal Plain, East and South Eastern Coastal Plain, North Eastern Ghat, Eastern Ghat High Land, South Eastern Ghat, Western Undulating Zone, Western Central Table Land, and Mid Central Table Land.

According to the 2011 Population Census, about 83.3% of Odisha's population lives in rural areas. Being an agrarian state, a significant portion of the rural labour force is engaged in agricultural activities. The agricultural sector is a major source of livelihood for many residents. In terms of Gross Value Added (GVA), the agricultural and allied sectors contributed approximately 21.27% in 2020–2021(A) and 21.38% in 2019–2020.

Odisha's total population stands at 419.74 lakhs, with 108.44 lakhs engaged in agriculture (cultivators and agricultural labourers). Among them, 41.04 lakh are cultivators, including 7.29 lakh female cultivators, who make a significant contribution to the state's agriculture. The climatic and soil conditions greatly influence agricultural practices in Odisha. The state experiences a tropical climate characterized by high temperatures, high humidity, moderate to heavy rainfall, and short, mild winters.

The average annual precipitation in Odisha is 1,451 millimetres. The state is highly susceptible to climate change impacts, largely due to its location on India's eastern coast and its 480 km long vulnerable coastline. Consequently, Odisha experiences recurring climate hazards such as cyclones and coastal erosion, with a series of climate-induced natural calamities affecting the region over the years (refer to Table 4.1).

The Kharif season is the primary cropping period, during which rice is the predominant crop, cultivated on approximately two-thirds of the cultivated land. Crop cultivation during the Rabi season is primarily limited to areas with access to irrigation facilities and residual moisture. Odisha cultivates various significant crops, including pulses (arhar, moong, biri, and kulthi), oilseeds (groundnut, sesamum, mustard, and niger), fibres (jute, mesta, and cotton), sugarcane, as well as vegetables and spices. During the Kharif season, rice dominates the cropped area, while during the Rabi season, pulses occupy nearly half of the cropped area. Additionally, a significant portion is occupied by vegetables, fibres, maize, and ragi (Agriculture Statistics 2018, GoO).

4.2.2 Description of the Study Areas

Data collection for this study was conducted in two extreme climatic regions: inland and coastal districts of Odisha. The geographical maps of the study districts are shown in Figure 4.1. The study covers three districts: Kendrapara, Mayurbhanj, and Balangir. Odisha's climate varies considerably, with droughts, floods, and cyclones regularly impacting the regional economy. Detailed descriptions of the study areas are provided below.

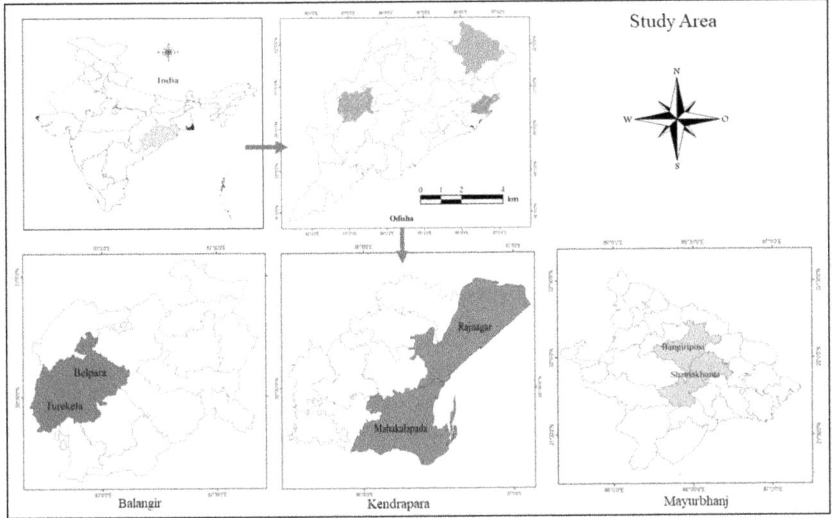

Figure 4.1 Study Area (Showing the Three Climate Vulnerable Districts as Study District).

4.2.2.1   Balangir District

Balangir District, an inland district located in the western part of Odisha, falls under the Western Central Tabled Land agro-climatic zone. It is situated between 20°9′ and 21°05′ North latitude and 82°41′ and 83°42′ East longitude. The district is bordered by Bargarh to the north, Kalahandi to the south, Subarnapur to the east, and Nuapada to the west. The climate is hot and humid, with droughts occurring more than thirteen times in the past three decades, making it a drought hotspot. About 10.9% of the cultivated land is irrigated during the Kharif season, and 5% during the Rabi season.

The main source of rainfall is the southwest monsoon, with the district receiving an average annual rainfall of 1,229.47 mm, which is below the state average. Approximately 80% of the total rainfall occurs between June and September. Droughts are frequent in this area.

4.2.2.2   Kendrapara District

Kendrapara District is situated in the eastern region of Odisha, with geographical coordinates between 86°14′ to 87°3′ East longitude and 20°21′ to 20°47′ North latitude. It is bordered by Bhadrak to the north, Jagatsinghapur to the south, the Bay of Bengal to the east, and Cuttack to the west. The climate is typically hot and humid in April and May, and cold in December and January. The monsoon season usually concludes in June. In 2018, the district received 1,885.8 mm of rain, significantly higher than the state average of 1,556 mm.

Approximately 77% of the cropped area is rainfed, while the remaining 23% is irrigated. Over the last two decades, Kendrapara has experienced six cyclones, the highest among all coastal districts in India. The district faces high risks of sea erosion and soil salinization (Maharjan, 2018). While high-altitude cropped land lacks adequate irrigation facilities, lower-altitude land suffers from water-logging due to insufficient drainage systems. Major crops produced in the district include paddy, black gram, green gram, sunflower, peanuts, and jute.

4.2.2.3 Mayurbhanj District

Mayurbhanj district is a land-locked district in the north-central plateau agro-climatic zone located in Northern Odisha. It is bordered by the states of West Bengal to the north, Jharkhand to the south, and districts such as Balasore to the west, and Keonjhar to the east. The district has a high density of tribal communities, accounting for 58% of its population, and features a tropical to sub-tropical climate with hot summer spells. Typically, the weather is hot and humid from July to September and cold from December to January. The monsoon rains usually arrive in June. In 2018, the district received 1,654.3 mm of rain, which was slightly below the state average of 1,556 mm.

Farming in Mayurbhanj District is predominantly rainfed, with 15% of the cultivable area irrigated during the Kharif season and 5.3% during the Rabi season. The main crops grown in the district include paddy, pulses, and oil-seeds. However, due to irregular rainfall patterns, there has been an increase in the area under pulses, oilseeds, and other cereals, while the area under Kharif paddy has decreased (Figure 4.2).

The temperature and rainfall trends of the study area are depicted in Figures 4.3 and 4.4. The analysis shows a consistent increase in the annual average temperature from 1982 to 2019, with the inland district experiencing a more pronounced rise compared to other parts of the study area. Conversely, the trend in average annual rainfall from 1988 to 2022 indicates a declining pattern specifically in the inland district of Balangir, while rainfall trends for the

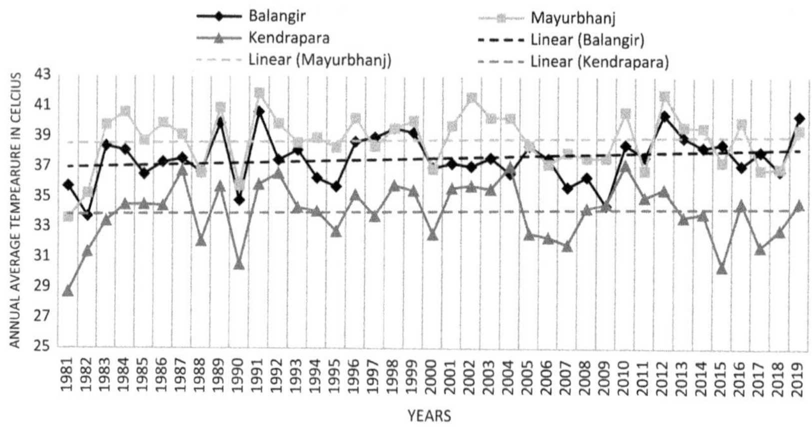

Figure 4.2 The Trend of Average Annual Temperature between 1988 and 2022 in Three Study Districts of Odisha.

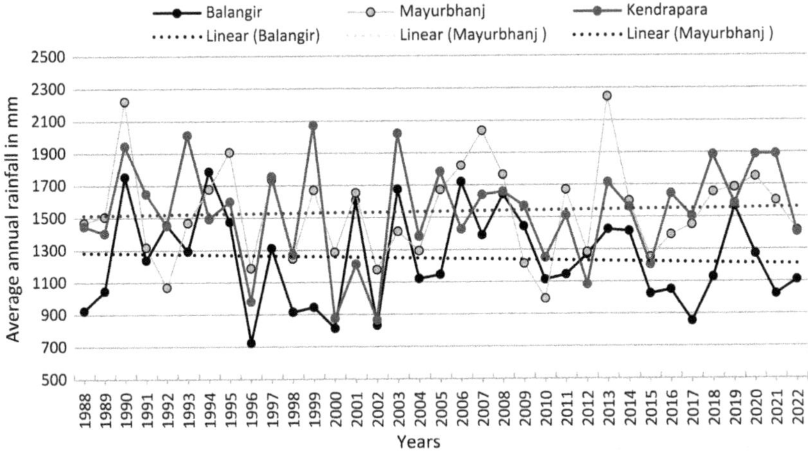

Figure 4.3 The Trend of Average Annual Rainfall between 1981 and 2019 in Three Study Districts of Odisha.

other two districts remain stable. These findings underscore the substantial impact of climate change on the study area, highlighting challenges related to agricultural production and water availability due to rising temperatures and shifting rainfall patterns.

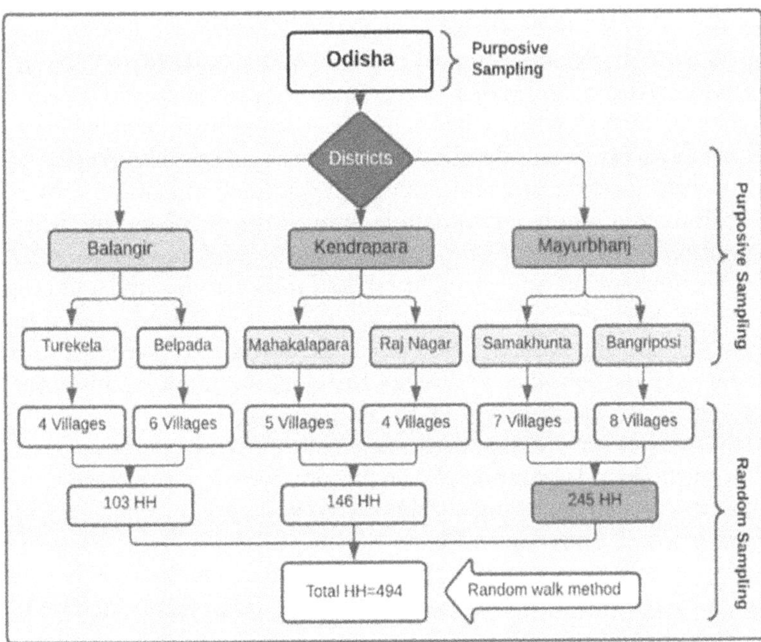

Figure 4.4 Study Design.

4.2.3 Sampling Determination

The current study utilizes cross-sectional data collected from 494 rural farmers in Odisha during the 2019–2020 production year. The sample size was determined using the statistical formula proposed by Arkin and Colton (1963):

$$n = \frac{NZ^2 * p * (1-p)}{Nd^2 + Z^2 * p(1-p)},$$

where,

n = required sample size (385)
N = total number of households (4,209,660)
Z = confidence level (at 95% level Z = 1.96)
p = estimated population proportion (0.5, this maximizes the sample size)
d = error limit of 5% (0.05)

Although the minimum required sample size was 385, data from 550 households were collected. After cleaning and preparation, 494 samples were used for the analysis.

4.2.4 Data Collection

The study employed a multistage stratified sampling approach. Initially, three districts from two different ecosystems in Odisha were selected to capture the heterogeneous effects of climate change and adoption patterns. In the second stage, blocks were purposively chosen based on their vulnerability to climate change, informed by reports and advice from agricultural officers and experts. Villages were then randomly selected from each block. Finally, households were chosen using a random walk method, where random numbers determined the number of paces between sample points and the direction of movement was decided by coin toss.

The survey questionnaire underwent pretesting and validation. Two experts in agriculture and resource economics reviewed the questionnaire, which was then pilot-tested to enhance validity and reliability. The questionnaire covered various sections including general household information, land and cropping patterns, income and yield details, perceptions of climate change, adoption of climate-smart agriculture (CSA) practices, access to government schemes (such as credit and subsidies), household asset information, input-output marketing challenges, and barriers to CSA adoption.

Additionally, farming characteristics such as farm size, cropping patterns, production practices, and utilization of CSA practices were queried. Government extension services such as training, subsidies, crop insurance, and credit accessibility were also assessed. Questions on farmers' perceptions of climate change regarding temperature increases, rainfall decreases, droughts, and floods were included.

Surveys were conducted in local languages (Odia, Sambalpuri) with the aid of trained enumerators and local guides. On average, each survey took between 45 minutes to one hour to complete. Responses were recorded directly on survey sheets, and informed consent was obtained from each household before the interview.

4.3 Description of the Variable

The adoption practices are identified from the pre-field visit and the farmer's field observation. Further, with consultation from Block Extension Officers, the selection of dependent variables has been fixed. We have included other adoption variables in our study; however, the popular and largely used by the local farmers are considered for the study analysis. The important CSA practices followed in the study areas are described in Table 4.2. The explanatory variables are constructed and given in Table 4.3.

Table 4.2 Definition of the CSA Practices or Dependent Variables

Adoption strategies	*As defined in this study*
Rescheduling planting	Due to uncertainty of the onset of monsoon, farmers do alter the planting dates. It is being rescheduled to prevent the delayed commencement and irregularity of the monsoon season. Sometimes farmers reschedule too early and sometimes late to the planting dates (Panda et al., 2013; Singh et al., 2018).
Crop rotation	Crop rotation is an agricultural technique involving the sequential cultivation of various crops in a field over time, aiming to improve soil quality and reduce pest and disease problems. This technique is widely employed in Odisha, particularly in the Balangir District. Its primary goals are to increase crop yield, maintain soil fertility, and reduce dependence on synthetic inputs. In accordance with the following sequence, the agricultural practice of crop rotation is implemented in this region. Pulses follow Paddy cultivation, then back to paddy, and subsequently followed by ginger and groundnut cultivation. (Abegunde et al., 2019).
Crop diversification	Crop diversification involves simultaneously cultivating multiple commodities on the same plot of land. Diversifying crops reduces the susceptibility of crops to pests and diseases, increases food security, and helps producers earn more money. Farmers in this region cultivate ginger and niger alongside pulses (Jha et al., 2018).
Soil conservation	Soil conservation refers to practices that prevent soil erosion and degradation, maintain or improve soil quality, and preserve soil fertility to guarantee sustainable crop production. Farmers practise terracing, afforestation, contour bunding, earth bunding, mulching and conservation tillage in the study region. Soil conservation also involves minimizing acidification, salinization, or other chemical contaminants to reduce erosional soil loss (Lobo et al., 2017; Singh et al., 2018).
Drought-resistant seeds	Drought-resistant seeds have been bred or genetically modified to grow successfully in arid conditions. These seeds can be planted in areas with insufficient water or irrigation, and they can still develop and produce a harvest despite the lack of water. In water-scarce regions and lack of irrigation areas, farmers plant drought-resistant seedlings of short-duration varieties or early maturity variety seeds (Khatri-Chhetri et al., 2016).

(Continued)

Table 4.2 (Continued)

Adoption strategies	As defined in this study
Agroforestry	Agroforestry is a land management technique that involves the cultivation of crops together with the planting of trees and shrubs. This method is also referred to as integrated crop and forest management. The aforementioned system exhibits multifunctionality in land use and holds the capacity to generate diverse benefits for the environment, society, and the economy. Agroforestry is a land management system that has been shown to mitigate soil erosion and enhance soil fertility (Jhariya et al., 2019).

Table 4.3 Explanatory Variables

Category	Variables	Expected outcome	Sources
Institutional	Govt. extension	+	(Bryan et al., 2013;
	Training	+	Jha et al., 2018;
	Farmers to farmer extn.	+	Azadi et al., 2019)
	Training	+	
	Machinery/seed subsidies	+	
	Cooperative society	+	
	Credit from a public bank	+	
Perception of climate change	Temperature is increasing	+	(Carlton et al., 2016;
	Rainfall is decreasing	+	Funk et al., 2020)
	Droughts and floods are increasing.	+	
	Experienced shocks	+	
Access to energy	Multiple/electricity/kerosine/diesel	+	(Jain et al., 2015; Ngigi et al., 2017)
Household Attributes	Age of the HH.	+/−	(Bryan et al., 2013;
	Education of the HH.	+	Abid et al., 2015;
	Years of farming	+	Tripathi and Mishra,
	Household size	+	2017; Musafiri
	Social category	+/−	et al., 2023)

Source: Compiled by the Authors.

4.4 Results and Discussion

Table 4.1 illustrates the adoption rates of various CSA practices among respondents in the study districts. It shows that 62% of respondents adopted soil conservation practices, while 59% implemented crop rotation. Seasonal or annual crop rotation is integral to CSA as it helps in preserving soil health, managing pests and weeds, and maintaining soil organic matter.

A higher proportion of farmers, approximately 65%, practised crop rotation in the inland districts compared to 48% in the coastal district. This disparity can be attributed to the higher-than-average rainfall experienced in coastal areas of eastern India, where farmers typically cultivate paddy throughout the year. In contrast, inland districts are more prone to drought due to inadequate rainfall,

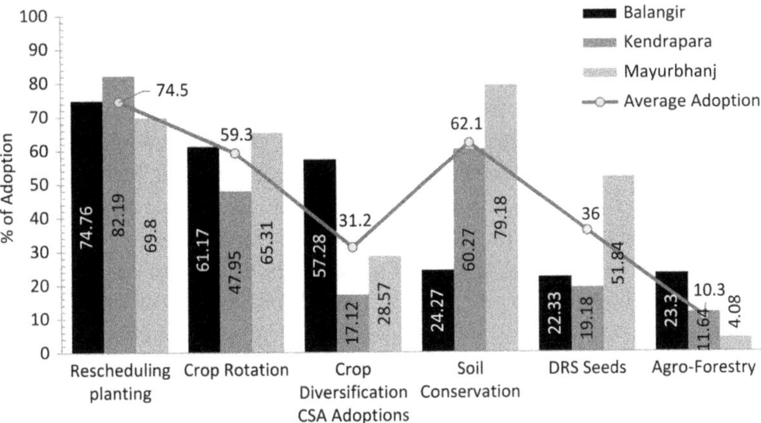

Figure 4.5 Adoption of CSA Practices across the District.

prompting farmers to adopt crop rotation strategies. These farmers commonly employ crop rotation systems such as rice-vegetable, rice-oilseeds, maize-pulses or oilseeds, and fibre-pulses on an annual basis. Respondents also adopted soil conservation measures such as gypsum application, increased field bund heights, and mulching. Additionally, they planted scrubs and constructed stone bunds along farm plot perimeters to mitigate soil erosion (Figure 4.5).

In the Mayurbhanj district, farmers predominantly cultivate five main crops: paddy, black gram, horse gram, green gram, and maize. The adoption of soil conservation practices is notable among respondents, including gypsum application, enhanced field bund heights, mulching, and crop rotation. Farmers in this district also plant scrubs and construct stone bunds along farm plot perimeters to mitigate soil erosion. Mayurbhanj district shows the highest adoption rates for soil conservation practices, followed by Kendrapara and Balangir districts.

To mitigate drought conditions, farmers in the study area have adopted short-duration variety seeds, also known as early maturity variety seeds. Approximately 36% of farmers used drought-resistant seeds on average. Varieties such as Swarna Sub-1, MTU-1010, Lalat, and Konark are used in medium-altitude lands, while Khandagiri, Heera, Kalinga-III, and Vandana varieties are preferred in high-altitude lands. In low-altitude lands, varieties like Swarna, Sub-1, CR-1014, Durga, Sarala, SR-10, Sonamani, and Lunishree are commonly adopted.

Agroforestry practices are followed by 10.3% of respondents across the study districts, with higher adoption rates in Balangir (23.3%), followed by Kendrapara (11.64%), and Mayurbhanj (4.08%) districts. Trees such as mango, cashew, guava, teak, and eucalyptus are planted on uplands, bunds, and ditches. Coconut-based agroforestry is predominant in Kendrapara district, while Bambusa nutans trees are primarily found in Mayurbhanj district.

The state government provides subsidies for machinery, seeds, and fertilizers. Farmers receive a 50% subsidy for tillers and 40% for tractors in Odisha, disbursed through Direct Bank Transfer (DBT). However, farmers often face challenges such as mistimed supply of subsidized seeds, which are sometimes unavailable when needed urgently or supplied too late after the planting season.

Regarding financial support, 28% of respondents received machinery subsidies, while 40% benefitted from seed subsidies. Agricultural credit is accessed by 46% of farmers through primary agriculture cooperative societies (PACS) and 24% from public-owned banks like the State Bank of India and Canara Bank. Regional rural banks like Utkal Grameen Bank also play a significant role in providing credit, typically secured against land records.

Farmers travel an average of 6.2 km to reach input markets from their homesteads and 13 km to reach the nearest agricultural extension office, known as the "block agriculture office". These offices provide crop advisories, access to agricultural schemes, and training sessions. Energy availability near agricultural fields is crucial for better adoption rates. Approximately 17.2% of farmers use multiple energy sources for irrigation, while 43% lack any energy source for irrigation purposes. Diesel is used by 14% of farmers for irrigation, and 22.5% have access to electricity near their agricultural fields (Figure 4.6).

Access to extension services plays a crucial role in agricultural practices and productivity. On average, farmers travel 6.2 km to reach input markets from their homesteads and 13 km to access the nearest agricultural extension office, known as the "block agriculture office". These extension offices are essential for farmers to receive crop advisories, access various agricultural schemes, and participate in training sessions conducted by extension officers. Energy availability near agricultural fields significantly influences adoption rates of agricultural practices. Due to the lack of surface irrigation facilities and unreliable rainfall, farmers largely depend on groundwater and local water bodies for irrigation. Approximately 17.2% of farmers utilize multiple energy sources for irrigation purposes, while 43% do not have any energy source available for irrigation. Among those who do, 14% use diesel for irrigation, and 22.5% have access to electricity near their agricultural fields (Tables 4.4 and 4.5).

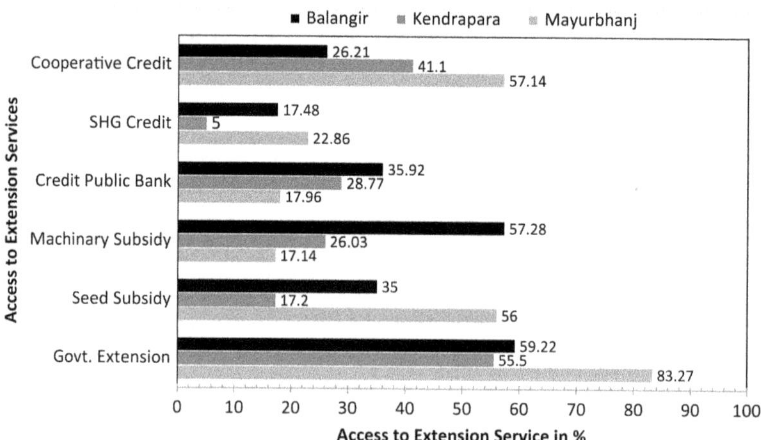

Figure 4.6 Access to Extensions Services.

Table 4.4 Descriptive Statistics

Variable	Variable description	Mean	Std. Dev.	Min	Max
Crop Rotation	Adoption of crop rotation (dummy, yes = 1, no = 0)	0.593	0.493	0	1
Integrated soil Management	Adoption of ISM (dummy, yes = 1, no = 0)	0.621	0.485	0	1
Total Area	Farmland size (in acre)	2.876	2.218	0	8.75
Farming experience	Household head farming experience (in numbers)	24.314	11.519	2	55
HH size	Number of family members (cont.)	4.891	1.655	1	11
Age	Age of farming household head	50.735	11.718	18	82
Govt. Extn.	Access to govt. extension (dummy, yes = 1, no = 0)	0.662		0	1
Farmers to farmer extn.	Contact with peer farmers (dummy, yes = 1, no = 0)	0.52		0	1
Training	If farmer gets training (dummy, yes = 1, no = 0)	0.312		0	1
Migration	If farmers migrate for work (dummy, yes = 1, no = 0)	0.261		0	1
Education	Years of schooling (cont.)	7.771	5.358	0	17
Machinery subsidies	If farmer avails machinery subsidy (dummy, yes = 1, no = 0)	0.281		0	1
Seeds subsidies	If farmer avails seeds subsidy (dummy, yes = 1, no = 0)	0.397		0	1
Cooperative credit	If farmer avails credit from cooperative society (dummy, yes = 1, no = 0)	0.46		0	1
Credit from Public Bank	If farmer avails credit from public banks (dummy, yes = 1, no = 0)	0.249		0	1
Perception _temp	If farmers perceive to increase in temperature (dummy, yes = 1, no = 0)	0.846		0	1
Perception _rainfall	If farmers perceive to decrease in temperature (dummy, yes = 1, no = 0)	0.769		0	1
Experienced Drought Shock	If farmer experiences drought (dummy, yes = 1, no = 0)	0.557		0	1
Experienced Flood Shock	If farmer experiences flood (dummy, yes = 1, no = 0)	0.255		0	1
Distance to Market	The distance of market from village in k.m. (cont.)	6.284	3.874	0	18
Multiple energy sources	If farmer uses multiple sources of energy in Agriculture (dummy, yes = 1, no = 0)	0.169		0	1
Access to Diesel	If farmer uses diesel for irrigation (dummy, yes = 1, no = 0)	0.138		0	1
Access to Electricity	If farmer uses electricity for irrigation (dummy, yes = 1, no = 0)	0.226		0	1
Livestock score 1	The index of various livestock	0	1.199	−1.249	9.406
Agri mechanization	The index of various machineries used in the agriculture	0	1.378	−1.164	5.629
Mayurbhanj	If farmer belongs to Mayurbhanj district (dummy, yes = 1, no = 0)	0.5	0.501	0	1
Balangir	If farmer belongs to Balangir district (dummy, yes = 1, no = 0)	0.209	0.407	0	1

Total sample size: 494

Table 4.5 Multivariate Probit Model Results

	Crop rotation	*ISM*
Total area cultivated	0.0357	0.0495
	(0.0354)	(0.0354)
Farming experience	−0.00319	−0.00712
	(0.00726)	(0.00740)
HH size	−0.0145	0.0363
	(0.0385)	(0.0391)
Age	−0.000206	0.000400
	(0.00728)	(0.00748)
Govt extn	0.493***	0.534***
	(0.133)	(0.132)
Farmers to farmer extn	−0.145	−0.00596
	(0.180)	(0.178)
Training	−0.0531	0.0998
	(0.146)	(0.147)
Migration	0.314**	0.0319
	(0.144)	(0.143)
Education	−0.00787	0.000118
	(0.0117)	(0.0118)
Machinery subsidies	0.167	0.0597
	(0.154)	(0.149)
Seeds subsidy	0.383***	−0.194
	(0.143)	(0.146)
Cooperative credit	0.156	0.140
	(0.132)	(0.134)
Credit from public banks	0.0309	0.285**
	(0.145)	(0.147)
Perception to increase temperature	−0.119	−0.155
	(0.194)	(0.199)
Perception to decrease rainfall	0.262	0.154
	(0.164)	(0.164)
Drought shock	−0.0837	0.0212
	(0.146)	(0.146)
Flood shock	−0.0181	0.0518
	(0.172)	(0.169)
Distance to market	−0.0117	−0.00852
	(0.0180)	(0.0178)
Multiple energy sources	0.211	0.189
	(0.179)	(0.184)
Access to diesel	0.626***	0.0472
	(0.215)	(0.210)
Access to electricity	0.506***	0.199
	(0.174)	(0.175)
Livestockscore_1	0.0682	0.0630
	(0.0571)	(0.0589)
Agr_mechanization index	0.0245	−0.0480
	(0.0550)	(0.0528)
Mayurbhanj	0.362	0.996***
	(0.262)	(0.265)
Balangir	−0.199	0.308
	(0.241)	(0.239)
_cons	−0.578	−0.914*
	(0.435)	(0.447)

(a) Standard errors in parentheses.
(b) *Significant at 10% level; **Significant at 5% level; ***Significant at 1% level.
(c) Observations: 494.
(d) Log likelihood = −563.1353, Wald chi2(50) = 144.96.

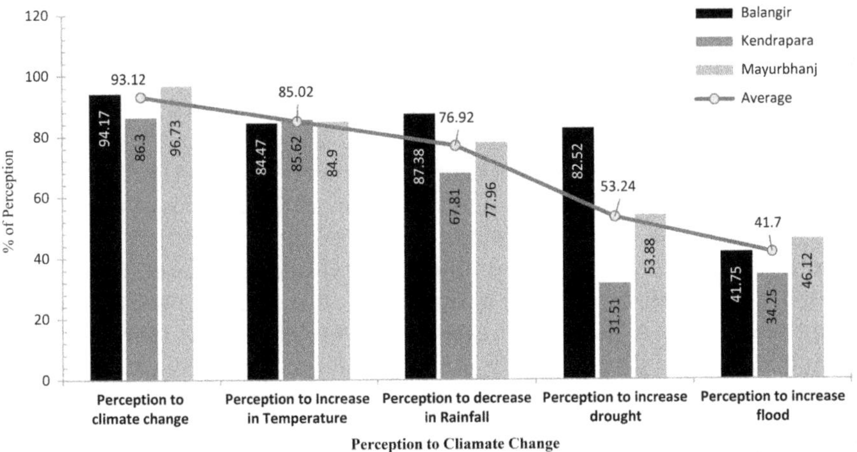

Figure 4.7 Perception to the Climate Change among the Rural Farmers.

The majority of the respondents (85%) perceive an increase in temperature, and (76%) perceive a decrease in rainfall due to climate change in the region. The respondents have adopted a range of CSA practices such as rescheduling planting (74.5%), crop rotation (59.3%), crop diversification (31.2%), soil conservation (62.1%), drought-resistant seeds (36%) and agroforestry (10.3%) to adapt to these weather anomalies (Figure 4.7).

4.5 The Broader Context

This chapter employed a multivariate probit model in which the findings of econometric modelling have been triangulated to explore the key determinants of the adoption of CSA practices. The key predictors are climate change perception, institutional support, and access to energy for irrigation that motivated farmers to adopt CSA practices. Those who have experienced climate shocks and believe that similar events may occur in the future are more likely to practice rescheduling planting, crop rotation, crop diversification, soil conservation, and drought-resistant seeds. Access to extension services is the main incentive for rural farmers to embrace CSA practices. Crop advisory and crop management training, demonstrations, farmers' field schools, machinery and seed subsidies, and access to institutional credit all help rural farmers improve their adaptability. Subsidies and government credit programmes are likely to enhance the adaptive capacity to adopt CSA practices. Agricultural input subsidies significantly impact the various adoption techniques used by farmers. Seed subsidies enable the adoption of crop rotation. Access to credit is a crucial factor in promoting the adoption of climate-smart agriculture (CSA) practices. The present study demonstrates a statistically significant positive effect of credit availability on the adoption of soil conservation practices, such as soil testing and bunding. These practices help to enhance soil fertility and preserve the top layer of the soil, which are critical components of sustainable agricultural production. Furthermore, access to inexpensive and affordable irrigation energy sources influences farmers' adoption behaviour positively. Subsidies from the government on power use and electricity infrastructure in farmers' fields can help to increase adoption.

The findings from this case study need to be discussed in the broader context of developing countries. Generally, rural farmers possess traditional knowledge about several adaptation practices to cope with the changing weather, pest disease, water run-off, and soil contamination. Since these problems are endemic in rural agricultural setup, farmers over the generations have practised cure and prevention measures to reduce the losses. For example, rescheduling the crop planting dates, either by planting early or late according to the rainfall prediction for the next season is a common adaptation practice among rural farmers. Similarly, crop rotation is also a proven practice traditionally followed by farmers to preserve soil health. This stock of traditional knowledge that emerged from generations of farming experiences in an evolving climate is a valuable source that can be used for policymaking in the agricultural sector. However, the current weather pattern observed in many parts of the globe is so erratic that it poses an unpresented level of confusion for the farmers. The traditional knowledge which has been built around a slowly evolving climate appears inadequate to deal with the vastly erratic and suddenly changing weather patterns that are currently observed. It is in this context that a modern approach to climate change adaptation is required, in which a mix of traditional and modern practices can be amalgamated to produce effective solutions for climate change threats.

Extension service will have to play a very decisive role in this effort to mitigate and adapt to climate change. During our qualitative data collection regarding expert and focus group discussion, we learned that a group of farmers from each village turned up for extension meetings and field demonstration training programmes. These farmers could be termed as "progressive farmers". During the early drives of extension meetings, these farmers showed interest in participating and became regular visitors to such extension programmes. They subsequently tried to train others in the village; however, certain personal traits such as innate abilities, risk preferences, and work ethics play a role in such peer-to-peer extension dissemination.

Farmers with continued access to extension services are more likely to integrate horticultural crops with traditional grains and pulses to diversify their farm income. Access to the agricultural extension variable used in this study measures how often a respondent engages with extension officers. The frequency of extension engagement decided the adoption rate of CSA practices. Several existing studies establish that factors such as government extension service and farmer field school participation are the major determinants of CSA adoption (Tanti et al., 2022). Further, Tanti and Jena (2023) found that inefficient extension services, a lack of coordination between farmers and technical staff, and a lack of timely subsidy disbursement creates obstacles to adopting sustainable agricultural practices. Hence, understanding the extension system's role in different geographical regions is crucial to prepare strategy on how to improve the existing extension systems.

Considering that agricultural extension service is a crucial element of farmers' support system in developing countries and the physical mode of extension support is not fully operational, the digital mode of extension service is gaining importance among the scientific community and policymakers. A significant penetration of smart mobile phones and internet access in rural areas has made it a viable option for reaching out to a larger pool of rural population. Digital extension if properly designed may reduce the cost of communication and transaction cost, increase market efficiency, promote economic development, and reduce poverty. Studies have established that digital extension enables a wide range of information dissemination, increases geographic coverage, and offers two-way communication (Aker and Mbiti, 2010; Aker, 2011; Rajkhowa and Qaim, 2021). Therefore, there is growing interest in literature evaluating the effect of digital extension services on agricultural development. One strand of the literature focuses on the effects of information access on rural farmers' performance and market efficiency through the usage of mobile phones and Internet services (Aker, 2011; Burga and Barreto, 2014). Digital extension service does not only focus on pricing efficiency but also production efficiency by enabling farmers' knowledge at various stages of crop production, this in turn enhances productivity, and income and supports economic development. The other studies in the literature primarily focus on whether digital extension supports to increase in farm productivity and income among rural households (Aker and Ksoll, 2016; Baumuller, 2018; Rajkhowa and Qaim, 2021). Rajkhowa and Qaim (2021) found that adopting personalized digital extension services is positively associated with improved agriculture performance (input intensity, productivity, and income). Contrastingly, Aker and Ksoll (2016) found that mobile phone coverage did not increase crop sales or farmgate prices in the Niger. In addition, Mitra et al. (2018) found that improving farmers' access to price information is unlikely to have positive outcomes on farmgate prices.

The present study has used the concept of "digital extension" based on the prevalent pattern of digital communication to farmers from government sources. The state government of Odisha has been implementing various agricultural subsidy schemes. The mobile phone numbers of the beneficiaries of these schemes exist in the government database. The government agricultural department sends SMS to these beneficiaries containing information about crop selection, planting dates, fertilizer dosage, and application procedure during the cultivation period. During the harvesting period, the prices of different crops are communicated to the farmers. Further, one-way call-based communication also takes place in which the farmers receive calls, and upon accepting the calls they receive information about inputs and outputs. Apart from this, in the farmers' field school and training sessions digital modes of communication such as audio-video clips and PowerPoint presentations are used. The agriculture-based non-government organizations (NGOs) make WhatsApp groups of those farmers who have smart mobile phones in their area of jurisdiction to send information on crops and inputs.

Note

1 Disclaimer: This chapter is a revised version of the open-access journal article authored by one of the authors of this book manuscript (Jena et al., 2023).

References

Abegunde, V. O., Sibanda, M. and Obi, A. (2019) 'Determinants of the adoption of climate-smart agricultural practices by small-scale farming households in King Cetshwayo District Municipality, South Africa', *Sustainability*, 12(1), p. 195.

Aker, J. C. (2011) 'Dial "a" for agriculture: A review of information and communication technologies for agricultural extension in developing countries', *Agricultural Economics*, 42(6), pp. 631–647.

Aker, J. C. and Ksoll, C. (2016) 'Can mobile phones improve agricultural outcomes? Evidence from a randomized experiment in Niger', *Food Policy*, 60, pp. 44–51. https://doi.org/10.1016/j.foodpol.2015.03.006

Aker, J. C. and Mbiti, I. M. (2010) 'Mobile phones and economic development in Africa', *Journal of Economic Perspectives*, 24, pp. 207–232.

Alem, Y., Eggert, H. and Ruhinduka, R. (2015) 'Improving welfare through climate-friendly agriculture: The case of the system of rice intensification', *Environmental & Resource Economics*, 62(2), pp. 243–263.

Arkin, H. and Colton, R. R. (1963) *Tables for statisticians*. New York: Barnes and Noble Inc.

Aryal, J. P., Farnworth, C. R., Khurana, R., Ray, S., Sapkota, T. B. and Rahut, D. B. (2020a) 'Does women's participation in agricultural technology adoption decisions affect the adoption of climate-smart agriculture? Insights from indo-gangetic plains of India', *Review of Development Economics*, 24(3), pp. 973–990.

Aryal, J. P., Sapkota, T. B., Khurana, R., Khatri-Chhetri, A., Rahut, D. B. and Jat, M. L. (2020b) 'Climate change and agriculture in South asia: Adaptation options in smallholder production systems', *Environment, Development and Sustainability*, 22(6), pp. 5045–5075.

Aryal, J. P., Sapkota, T. B., Rahut, D. B., Marenya, P. and Stirling, C. M. (2021) 'Climate risks and adaptation strategies of farmers in East Africa and South Asia', *Scientific Reports*, 11(1), p. 10489.

Azadi, Y., Yazdanpanah, M. and Mahmoudi, H. (2019) 'Understanding smallholder farmers' adaptation behaviors through climate change beliefs, risk perception, trust, and psychological distance: Evidence from wheat growers in Iran,' *Journal of Environmental Management*, 250, p. 109456.

Banerjee, R. R. (2014) 'Farmers' perception of climate change, impact and adaptation strategies: A case study of four Villages in the semi-arid regions of India', *Natural Hazards*, 75(3), pp. 2829–2845.

Banerjee, R., Kamanda, J., Bantilan, C. and Singh, N. P. (2013) 'Exploring the relationship between local institutions in SAT India and adaptation to climate variability', *Natural Hazards*, 65, pp. 1443–1464.

Baumuller, H. (2018) 'The little we know: An exploratory literature review on the utility of mobile phone-enabled services for smallholder farmers', *Journal of International Development*, 30(1), pp. 134–154. http://doi.wiley.com/10.1002/jid.3314.

Bhatta, G. D., Aggarwal, P. K., Shrivastava, A. K. and Sproule, L. (2016) 'Is rainfall gradient a factor of livelihood diversification? Empirical evidence from around climatic hotspots in indo-gangetic plains', *Environment, Development and Sustainability*, 18(6), pp. 1657–1678.

Branca, G., McCarthy, N., Lipper, L. and Jolejole, M. C. (2011) 'Climate-smart agriculture: A synthesis of empirical evidence of food security and mitigation benefits from improved cropland management', *Mitigation of Climate Change in Agriculture Series*, 3, pp. 1–42.

Bryan, E., Ringler, C., Okoba, B., Roncoli, C., Silvestri, S. and Herrero, M. (2013) 'Adapting agriculture to climate change in Kenya: Household strategies and determinants,' *Journal of Environmental Management*, 114, pp. 26–35.

Burga, R. and Barreto, M. E. G. (2014) *The effect of internet and cell phones on employment and agricultural production in rural Villages in Peru*. Universidad de Piura.

Carlton, J. S., Mase, A. S., Knutson, C. L., Lemos, M. C., Haigh, T., Todey, D. P. and Prokopy, L. S. (2016) 'The effects of extreme drought on climate change beliefs, risk perceptions, and adaptation attitudes,' *Climatic Change*, 135, pp. 211–226.

Das, S. (2016) 'Economics of natural disasters in Odisha,' in The economy of Odisha: A profile (pp. 266–300).

Das, U., Ansari, M. A. and Ghosh, S. (2022) 'Effectiveness and upscaling potential of climate-smart agriculture interventions: Farmers' participatory prioritization and livelihood indicators as its determinants', *Agricultural Systems*, 203, p. 103515.

Deressa, T. T., Hassan, R. M. and Ringler, C. (2011) 'Perception of and adaptation to climate change by farmers in the Nile basin of Ethiopia', *Journal of Agricultural Science*, 149(1), pp. 23–31.

Dougill, A. J., Whitfield, S., Stringer, L. C., Vincent, K., Wood, B. T., Chinseu, E. L. and Mkwambisi, D. D. (2017) 'Mainstreaming conservation agriculture in Malawi: Knowledge gaps and institutional barriers', *Journal of Environmental Management*, 195, pp. 25–34.

Fisher, A. C., Hanemann, W. M., Roberts, M. J. and Schlenker, W. (2012) 'The economic impacts of climate change: Evidence from agricultural output and random fluctuations in weather: Comment', *American Economic Review*, 102(7), pp. 3749–3760.

Funk, C., Sathyan, A. R., Winker, P. and Breuer, L. (2020) 'Changing climate-Changing livelihood: smallholder's perceptions and adaption strategies,' *Journal of Environmental Management*, 259, p. 109702.

Haq, S. U., Boz, I. and Shahbaz, P. (2021) 'Adoption of climate-smart agriculture practices and differentiated nutritional outcome among rural households: A case of Punjab province, Pakistan', *Food Security*, 13, pp. 913–931.

Hariharan, V. K., Mittal, S., Rai, M., Agarwal, T., Kalvaniya, K. C., Stirling, C. M. and Jat, M. L. (2020) 'Does climate-smart village approach influence gender equality in farming households? A case of two contrasting ecologies in India', *Climatic Change*, 158, pp. 77–90.

Harvey, C. A., Chacón, M., Donatti, C. I., Garen, E., Hannah, L., Andrade, A. and Wollenberg, E. (2014) 'Climate-smart landscapes: Opportunities and challenges for integrating adaptation and mitigation in tropical agriculture', *Conservation Letters*, 7(2), pp. 77–90.

Jain, A., Ray, S., Ganesan, K., Aklin, M., Cheng, C. Y. and Urpelainen, J. (2015) Access to clean cooking energy and electricity: Survey of states.

Jaleta, M., Kassie, M., Tesfaye, K., Teklewold, T., Jena, P. R., Marenya, P. and Erenstein, O. (2016) 'Resource saving and productivity enhancing impacts of crop management innovation packages in Ethiopia,' *Agricultural Economics*, 47(5), pp. 513–522.

Jena, P. R. (2019) 'Can minimum tillage enhance productivity? Evidence from small-holder farmers in Kenya', *Journal of Cleaner Production*, 218, pp. 465–475.

Jena, P. R., Tanti, P. C. and Maharjan, K. L. (2023) 'Determinants of adoption of climate resilient practices and their impact on yield and household income,' *Journal of Agriculture and Food Research*, 14, p. 100659.

Jha, C. K., Gupta, V., Chattopadhyay, U. and Amarayil Sreeraman, B. (2018) 'Migration as an adaptation strategy to cope with climate change: A study of farmers' migration in rural India', *International Journal of Climate Change Strategies and Management*, 10(1), pp. 121–141.

Jhariya, M. K., Banerjee, A., Yadav, D. K. and Raj, A. (2019) 'Agroforestry and climate change: Issues, challenges, and the way forward.' In *Agroforestry and climate change* (pp. 1–34). Apple Academic Press. https://www.taylorfrancis.com/chapters/edit/10.1201/9780429057274-1/agroforestry-climate-change-issues-challenges-way-forward-jhariya-banerjee-yadav-abhishek-raj.

Kalli, R. and Jena, P. R. (2020) 'Impact of climate change on crop yields: Evidence from irrigated and dry land cultivation in semi-arid region of India', *Journal of Environmental Accounting and Management*, 8(1), pp. 19–30.

Kalli, R. and Jena, P. R.. (2022) 'How large is the farm income loss due to climate change? Evidence from India', China Agricultural Economic Review.

Kangogo, D., Dentoni, D. and Bijman, J. (2021) 'Adoption of climate-smart agriculture among smallholder farmers: Does farmer entrepreneurship matter?', *Land Use Policy*, 109, p. 105666.

Khatri-Chhetri, A., Aryal, J. P., Sapkota, T. B. and Khurana, R. (2016) 'Economic benefits of climate-smart agricultural practices to smallholder farmers in the Indo-Gangetic Plains of India,' *Current* Science, pp. 1251–1256.

Lobo, C., Chattopadhyay, N. and Rao, K. V. (2017) 'Making smallholder farming climate-smart: Integrated agrometeorological services', *Economic and Political Weekly*, 52(1), pp. 53–58.

Maharjan, S. (2018) 'Adoption of multiple climate-smart agricultural practices in the gangetic plains of Bihar, India', *International Journal of Climate Change Strategies and Management*, 10(3), pp. 407–427.

Makate, C., Makate, M. and Mango, N. (2019) 'Wealth-related inequalities in the adoption of drought-tolerant maize and conservation agriculture in Zimbabwe', *Food Security*, 11(4), pp. 881–896.

Mashi, S. A., Inkani, A. I. and Obaro, D. O. (2022) 'Determinants of awareness levels of climate-smart agricultural technologies and practices of urban farmers in Kuje, Abuja, Nigeria', *Technology in Society*, 70, p. 102030.

Mgendi, G., Mao, S. and Qiao, F. (2022) 'Does agricultural training and demonstration matter in technology adoption? The empirical evidence from small rice farmers in Tanzania', *Technology in Society*, 70, p. 102024.

Mishra, D., Sahu, N. C. and Sahoo, D. (2016) 'Impact of climate change on agricultural production of Odisha (India): A Ricardian analysis', '*Regional Environmental Change*, 16(2), pp. 575–584.

Mitra, S., Mookherjee, D., Torero, M. and Visaria, S. (2018) 'Asymmetric information and middleman margins: An experiment with Indian potato farmers', *Review of Economics and Statistics*, 100(1), pp. 1–13.

Mizik, T. (2021) 'Climate-smart agriculture on small-scale farms: A systematic literature review,' *Agronomy*, 11(6), p. 1096.

Mohanty, B. B. and Lenka, P. K. (2019) 'Small farmers' suicide in Odisha', *Economic and Political Weekly*, 54(22), p. 51.

Mujeyi, A., Mudhara, M. and Mutenje, M. (2021) 'The impact of climate smart agriculture on household welfare in smallholder integrated crop–livestock farming systems: evidence from Zimbabwe,' *Agriculture & Food Security*, 10, pp. 1–15.

Musafiri, C. M., Kiboi, M., Ng'etich, O. K., Okoti, M., Kosgei, D. K. and Ngetich, F. K. (2023) 'Carbon footprint of smallholder rain-fed sorghum cropping systems of Kenya: A typology-based approach,' *Cleaner and Circular Bioeconomy*, 6, p. 100060.

Ngigi, M. W., Mueller, U. and Birner, R. (2017) 'Gender differences in climate change adaptation strategies and participation in group-based approaches: An intra-household analysis from rural Kenya,' *Ecological Economics*, 138, pp. 99–108.

Osumba, J. J., Recha, J. W. and Oroma, G. W. (2021) 'Transforming agricultural extension service delivery through innovative bottom–up climate-resilient agribusiness farmer field schools', *Sustainability*, 13(7), p. 3938.

Panda, A. (2016) 'Exploring climate change perceptions, rainfall trends and perceived barriers to adaptation in a drought affected region in India', *Natural Hazards*, 84, pp. 777–796.

Panda, A., Sharma, U., Ninan, K. and Patt, A. (2013) 'Adaptive capacity contributing to improved agricultural productivity at the household level: Empirical findings highlighting the importance of crop insurance', *Global Environmental Change*, 23, pp. 782–790.

Pattanayak, U. and Mallick, M.. (2016) 'Incidence of farmers indebtedness and suicide in Odisha: An analysis.' *International Journal of Multidisciplinary Approach and Studies*, 3(3), pp. 1–12.

Rajkhowa, P. and Qaim, M. (2021) 'Personalized digital extension services and agricultural performance: Evidence from smallholder farmers in India', *PloS One*, 16(10), p. e0259319.

Sahoo, A. and Rath, N. (2023) 'Assessment of climate change, water poverty and risk communities: Some insights from Western Odisha'. *GeoJournal*, 88, 1–14.

Sahu, N. C. and Mishra, D. (2013) 'Analysis of perception and adaptability strategies of the farmers to climate change in Odisha, India', *APCBEE Procedia*, 5, pp. 123–127.

Sain, G., Loboguerrero, A. M., Corner-Dolloff, C., Lizarazo, M., Nowak, A., Martínez-Barón, D. and Andrieu, N. (2017) 'Costs and benefits of climate-smart agriculture: The case of the dry corridor in Guatemala', *Agricultural Systems*, 151, pp. 163–173.

Singh, N. P., Anand, B. and Khan, M. A. (2018) 'Microlevel perception to climate change and adaptation issues: A prelude to mainstreaming climate adaptation into the developmental landscape in India', *Natural Hazards*, 92(3), pp. 1287–1304.

Tanti, P. C., Jena, P. R., Aryal, J. P. and Rahut, D. B. (2022) 'Role of institutional factors in climate-smart technology adoption in agriculture: Evidence from an Eastern Indian state,' *Environ Challenges*, 7, p. 100498.

Tanti, P. C. and Jena, P. R. (2023) 'Perception on climate change, access to extension service and energy sources determining adoption of climate-smart practices: A multivariate approach', *Journal of Arid Environments*, 212, p. 104961.

The Hindu (2022) India suffered income loss of $159 billion in key sectors due to extreme heat in 2021: Report. Retrieved from https://www.thehindu.com/business/Economy/india-suffered-income-loss-of-159-billion-in-key-sectors-due-to-extreme-heat-in-2021-report/article66035523.ece

Tripathi, A. and Mishra, A. K. (2017) 'Knowledge and passive adaptation to climate change: An example from Indian farmers', *Climate Risk Management*, 16, pp. 195–207.

Zakaria, A., Azumah, S. B., Appiah-Twumasi, M. and Dagunga, G. (2020) 'Adoption of climate-smart agricultural practices among farm households in Ghana: The role of farmer participation in training programmes', *Technology in Society*, 63, p. 101338.

5 Role of Gender in Adopting Climate Smart Agriculture Practices

5.1 Women at the Forefront of Climate Effect

The observable consequences of climate change are progressively more conspicuous in contemporary times. These include the difference in the mean temperature and precipitation levels, modifications in the magnitude, timing, and spatial spread of precipitation, a surge in the frequency of severe climate-induced events such as droughts and floods, and a rise in sea levels (IPCC, 2007). Anthropogenic climate change poses the greatest threat to developing nations. Rural farmers in developing nations are highly vulnerable to extreme events. As a result of the phenomenon of climate change, female farmers encounter a greater number of obstacles in comparison to their male counterparts. Gender disparities persist in the fight against climate change in developing countries. Women constitute 70% of the population living below the poverty line in the developing world, which amounts to 1.3 billion individuals. Women contribute to agriculture significantly and account for about 43% of the global agricultural labour force (FAO, 2011). The threats posed by global warming have not sufficiently convinced policymakers of the significance of positioning women at the centre of their vision for sustainable development (Denton, 2002). Still, research into climate change's many gender-related aspects is severely lacking (MacGregor, 2010).

The presence of structural inequality and disempowerment has a detrimental impact on the capacity of women farmers to address issues related to climate change and food security effectively (Brody et al., 2008). The phenomenon of climate change is anticipated to have a disproportionate impact on female smallholder farmers, as it is likely to perpetuate pre-existing gender disparities and amplify the socioeconomic and political hazards that these farmers encounter (Tschakert and Machado, 2012; Steinfield and Holt, 2020; Tantoh et al., 2021).

Rural women farmers are more sensitive towards climate change due to their work profile which includes childcare, gathering firewood, collecting drinking water, managing most of the household work, and also executing most of the agricultural operations as farm labour (Nellemann et al., 2011; Goh, 2012). The responsibility of women increases when their male counterpart migrates

DOI: 10.4324/9781003290445-5

to another city in search of work. The exposure to climate change increases for a woman farmer when she takes responsibility for farming activities without a male (FAO, 2011). Further, the adaptive capacity is constrained by the factors such as less access to agricultural resources such as land, extension services, inputs, and mechanization. The social norms and gender roles in society are also an obstacle to the adaptive capacity of women farmers (Kakota et al., 2011).

The issue of climate change presents obstacles in the efforts to sustain and enhance female smallholder farmers' agricultural and labour productivity (Murray et al., 2016). Climate change adversely impacts the socioeconomic life of women. It affects the health condition likely to increase cardiovascular and respiratory diseases (Denton, 2002). FAO (2011) has examined the impact of climate change and differential risk, focusing on the gender dimensions. The report finds a strong correlation between gender equality and climate change, particularly in rural agricultural contexts where women face restricted access to clean energy, technical support for crop irrigation, and insufficient low-cost farming inputs and credit from financial institutions. This review explores the role of women in adopting and upscaling climate smart agriculture (CSA) practices worldwide.

The discourse surrounding the gender aspects of climate change and food security has been a topic of discussion in the field of development literature and practice for a considerable period. The inadequate comprehension of the intersection between climate change and socioeconomic challenges faced by susceptible populations in regions prone to conflict is a current issue (Omolo, 2010). Female farmers may experience a lack of autonomy compared to their male counterparts when making significant decisions regarding improvements in agricultural practices (Murray et al., 2016). Women's labour in agriculture is subject to limitations due to their involvement in the unpaid care economy, which may fluctuate throughout their lifespan, such as before childbirth, during childcare, or while caring for older people (Peterman et al., 2014). Various studies have asserted that women smallholder farmers must have access to relevant inputs and tools to take up CSA practices. The adaptability among the women head farmers is low. It is due to financial or resource limitations in a patriarchal society dominated by men, a lack of access to information and extension services, and a greater labour demand for women head farmers (Deressa et al., 2009; Bryan et al., 2013; Jost et al., 2016). Despite efforts to promote the adoption of new labour-productivity-enhancing technologies and Community Supported Agriculture (CSA) practices, women farmers still encounter obstacles (UN-Women, 2015).

5.2　Evidence on Effects of Climate Change on Gender

The review of the gender analysis can provide insights into the decision-making processes concerning climate-smart agricultural practices, such as intra-household bargaining and resource allocation. Various groups exist in a household, such as households headed by men or women or women residing in households headed by men (Murray et al., 2016). Generally, ideological, social, and economic dominancy and encounters exist in making a decision which is a limitation towards the progressive adoption decision. Firstly, the review tries to explore the combination of structural factors that give rise to vulnerability, coupled with the risk posed by climate change, which contributes to widening the disparity in agricultural resources available to female farmers. Secondly, the review tries to identify gender roles in the adoption of CSA practices in developing countries.

Demetriades and Esplen (2010) described that the persistent gender inequality in society and climate change could enhance the poverty rate of women more than that of men. The intra-household disparity in the distribution of power and property and less access to and control over financial, material, and human resources undermines the capabilities to adapt to predicted or existing climate change issues. The statutory and customary laws prevent women from accessing their land rights, which could lead to inaccessibility to credit. The biased nature of various institutions treats men only as farmers and debars women from accessing the extension and technology in agriculture in a climate-vulnerable region. Lack of access to land rights, credit, and technology among women farmers could be a significant constraint to adapting to climate change. Women also face sexual and domestic violence after climate change-induced disasters (Bartlett, 2008).

Jury (2010) conducted a case study in Namibia and tried to find the relationship between gender and climate change. The author asserted that the impact of climate change on the means of subsistence of individuals varies based on gender. This research suggests that households headed by widows or bachelors experience greater difficulty managing the risks and vulnerabilities associated with climatic variability and changes than those headed by females. The level of impact that climate change has on the environment and the extent to which women are exposed to climatic risks are major.

Ribeiro and Chaúque (2010) studied Mozambique and found that both men and women were impacted by climate change, but the degree of impact was different for both genders. Women farmers had access to resources but did not have control and property rights over them. The community had experienced consecutive droughts over the past two years, resulting in a rise in male individuals migrating to other locations in pursuit of employment opportunities. Consequently, the involvement of women in productive agricultural work has significantly escalated. An article by Glazebrook (2011) examines the distinct influence of climate change on women residing in the northeastern region of Ghana. The argument posited by the author is that women residing in rural areas are especially susceptible to the impacts of climate change, owing to their significant dependence on agriculture as a source of sustenance and their limited access to resources and information necessary for adapting to fluctuating weather conditions.

Babugura et al. (2010) present a case study that examines the effects of climate change on gender in South Africa. The research reveals that climate change disproportionately impacts women, as they are more susceptible to its economic, social, and health consequences owing to pre-existing gender disparities. Integrating gender perspectives into climate-related policies, enhancing women's involvement in decision-making processes, and furnishing women with economic prospects in sustainable sectors help to face the climate change issue among female farmers. Arora-Jonsson (2011) examines the different lenses employed while discussing women, gender, and climate change. The author argues that the account mentioned above fails to acknowledge women's susceptibility to the impacts of climate change and their potential to contribute to adaptive measures. The article underscores the necessity of transcending facile generalizations and adopting a more comprehensive and intersectional strategy towards climate action.

Arora-Jonsson (2011) indicated that women are disproportionately affected by climate disasters due to their socially constructed gender roles and responsibilities and their comparatively disadvantaged economic status, particularly in developing nations. Gender disparities in Bangladesh regarding the exercise of human rights, political and economic standing, land possession, housing circumstances, susceptibility to violence, education, and health (specifically reproductive and sexual health) render women more susceptible to the effects of climate change-triggered calamities, both before and after of their occurrence.

Onta and Resurreccion (2011) found the intersections of gender and caste in the context of climate change adoption. They have presented novel evidence of potential unintended consequences and opportunities and resilient practices related to gender and caste in Nepal. The interdependence and interconnectivity of various factors play a crucial role in devising effective adoption strategies, particularly in light of the escalating issue of food insecurity caused by reduced crop productivity in the face of climate change.

Goh's (2012) review finds the gender-specific effects of climate change on individuals residing in developing nations. Women and men are impacted differently. The impact of climate-related incidents on the well-being and assets of men and women varies. Climate-related shocks disproportionately impact women in comparison to men. The adverse effects of climate change on agricultural production, food security, health, water and energy resources, climate-induced migration and conflict, and climate-related natural disasters exhibit gender-based variations, leading to differential impacts on the assets and well-being of both genders.

Alston (2013) compiles research from various regions, including Australia, Canada, Africa, Asia, and Europe, to showcase the growing body of evidence indicating the significance of gender on climate change. The effects of climate change are experienced differently by women and men, both during and after a climate crisis. These experiences are influenced by a range of factors, including cultural norms and practices, work roles and access to resources, safety and security, and varying levels of vulnerability resulting from the interplay of these factors. Lambrou and Nelson (2013) confirm that gender plays a significant role in how farmers experience and express climate variability in their coping strategies to guarantee their livelihoods and food security. As they pursue food security, women's and men's perceptions and responses to the effects of altered climatic conditions differ significantly.

Carr and Thompson (2014) acknowledge gender and climate change adoption in rural contexts showing that men and women often perceive climate unpredictability and change differently. Evidence shows that women cultivate crops with different biophysical properties than men in rural contexts. Women also have limited decision-making and access to land and input. Women are also developing locally suitable adoptions to climate change.

Sugden et al. (2014) reported that in the context of the Eastern Gangetic Plain and agrarian stress, male out-migration is the main cause of gendered vulnerability. This has changed how to work and resources are shared. Even though people from all walks of life move, the most vulnerable are women from poor farmers and tenant families. Women are coming up with ways to deal with the effects of climate change, building skills and plans that work well in their areas.

Alston (2015) in his book noted that women face greater vulnerability during and after a climate-induced disaster. Women are more likely to die, suffer, and face violence due to less control over resources and the inequality structure of society. Women have less access to resources that are needed to prepare for disaster, adoption, and mitigation. Yadav and Lal (2018) noted women farmers' exposure to climate change. The susceptibility of women populations to climate change can be attributed to various factors such as poverty, gender inequality, insecure land rights, high dependence on agriculture, and limited access to education and information. The susceptibility is further complicated by the limited resources, societal marginalization, restricted mobility, and absence of involvement in the disaster response decision-making procedures.

Ylipaa et al. (2019) examine how men and women in Thái Bnh adapt to climate change and make a living as farmers. The study finds that male and female farmers have different rights and responsibilities. This leads to unequal opportunities and immobility for women, which makes them more vulnerable to the effects of climate change and threatens to reduce their ability to adapt to climate change. Women do not have rights or control over things they are responsible for, but they do not have rights or control over them. So, women cannot get access to and add to knowledge production to help them escape their difficult situation. They also do not have the power to participate in or change policymaking.

The study by Balikoowa et al. (2019) asserts that female-headed households are more vulnerable than male-headed households. Women enjoy fewer assets and opportunities compared to men. Women farmers usually face resource constraints when a female in the household is undertaking the headship without male members. Due to resource constraints, the adaptive capacity is decreasing among women farmers. Households headed by females exhibited less resilient livelihoods, elevating their susceptibility to the impacts of climate change. Female-headed households, primarily consisting of widowed women, possessed a smaller amount of land too.

Alhassan et al. (2018) found the vulnerability to climate change among male head and female farmers in Ghana. The vulnerability of female-headed households was found to be significantly higher in relation to their socio-demographic profile, livelihood strategies, social network, and access to water and food than male-headed households. Female-headed households exhibit greater sensitivity to climate change and variability, rendering them comparatively more vulnerable in terms of adaptive capacity when compared to male-headed households. Overall, households headed by females exhibited greater susceptibility to climate change impacts and variability than those headed by males.

Asadullah and Kambhampati (2021) assert that women's labour participation in agriculture gradually decreases in the developing world, where female farmers' participation is increasing. However, labour participation in agriculture has not enhanced women's lives beyond being a farmer, although female participation in agriculture has enhanced food security and women's empowerment. The formation of Women in Agriculture (WIA) farmer groups and cooperative groups has facilitated the dissemination of agricultural innovations and provided women farmers with better access to farm inputs and credit than they would have as individuals (Ogunlela and Mukhtar, 2009). However, multiple studies reveal that women lag behind men in agricultural technology adoption on climate change (Rola-Rubzen et al., 2020; Vemireddy and Choudhary, 2021).

Majumder and Shah (2017) noted, in the context of Indian agriculture and household livelihood, that women play a pivotal role in providing essential support. Despite this, their participation is often limited to the role of workers. The current availability of farm tools is primarily geared towards male farmers, leaving rural women to rely on traditional tools and methods. This has decreased efficiency, increased physical strain, increased occupational health hazards, and reduced income for these women. Theis et al. (2018) did extensive qualitative studies in Bangladesh to show women do not adopt water-level adoption irrigation pumps due to technological complexity, difficulty in operating machines due to physical requirements, and difficulty in hiring and controlling the labour force. A study by Gouse et al. (2016) shows that female farmers put importance on labour-saving technology and pest control variety of seeds. The male farmers focus on yield benefits, whereas the female farmers prioritize the taste, quality, and simplicity of cultivation.

Khatri-Chhetri et al. (2020) developed a women's CSA adoption map to determine women producers' adoption of CSA practices and the resulting impact on drudgery. Nepal's agriculture is swiftly becoming increasingly female-dominated. The Nepalese women farmers have adopted direct seeded rice (zero tillage and minimal tillage using machinery), green manuring (GM), laser land levelling (LLL), and a rice intensification system (SRI). In addition to reducing women's agricultural drudgery, these practices have increased productivity and farm income.

Hariharan et al. (2020) compared women's participation in the climate-smart village (CSV) approach and looked at how it affected gender equality in Haryana and Bihar between 2012 and 2015. Both states examined showed an increase in women working in CSVs and in women's political, social, and agricultural lives across Haryana. The economic, social, and agricultural participation of Bihar's CSVs is also improving. To some extent, promoting change in gender equality helps women in the interviewed groups develop their skills and abilities. The analysis shows the benefits of CSA interventions in the CSV approach.

Balasha et al. (2023) examined gender differences in climate change perception and how they affect farmers' adoption of sustainable practices. About 50% of women farmers preferred Indigenous climate knowledge, while 61% of men farmers said experience and exchange helped them read and predict climate trends. Male farmers used less pesticide for climate change adoption than female farmers. More women (50%) than men (32%) planted living hedges to prevent field erosion. Nnadi et al. (2023) investigated the challenges and opportunities linked to gender in the region of South-East Nigeria. The study highlights significant disparities between male and female-headed households adopting migration and livelihood diversification strategies. Women faced challenges related to inadequate farming knowledge, while male farmers encountered difficulties due to insufficient knowledge of mechanization. The female demographic exhibited lower involvement rates in training programmes and limited availability of telecommunication resources conducive to hands-on learning. Similarly, the study of Mphande et al. (2022) examined the difference between male and female-headed households and found that female farmers encountered difficulties in obtaining hybrid legume seeds, inoculants, and promoting beans. One of the challenges faced by men in the soybean industry was the issue of low market prices.

Nchanji et al. (2022) show that different groups of men and women use and adopt climate-smart agriculture technologies for growing beans differently. Hove and Gweme (2018) found that adopting conservation farming among the women farmers of Zimbabwe enhanced food security in the region. Farmers who implemented conservation agriculture techniques while adhering to the recommended components and employing appropriate strategies experienced an enhancement in their food security during the dry season. Ge et al. (2023) examined the difference between males and females in adopting crop diversification as a CSA practice. Adopting crop diversification among male farmers depends significantly on their attitudes, subjective norms, behavioural control, and environmental knowledge. On the contrary, female farmers' adoption of crop diversification is not influenced significantly by their attitudes, subjective norms, perceived behavioural control, and environmental knowledge.

Rural women are crucial in managing natural resources, the environment, and agriculture production. Compared to men, women face multiple challenges related to financial and resource constraints and lower access to information and extension services (Tall et al., 2014; Huyer, 2016). In the previous section, we discussed why women are less adaptive than men. However, some agricultural economists believe women are active agents in devising responses to climate change and adapting to its effects. (Denton, 2002; Dankelman, 2010; Huyer, 2016). Edmunds et al. (2013) believe that women's participation in advancing technology and management decisions could improve the positive outcomes of community activities and lead to gender justice. Anderson et al. (2013) argue for incorporating leadership from rural and indigenous women to address water system infrastructure, quality, and security issues in rural regions. Female farmers have the potential to improve food security and increase their livelihood opportunities by expanding the cultivation of high-value crops (Quisumbing et al., 1996). Women engaged in agricultural activities play a crucial role in safeguarding biodiversity. The engagement of individuals in crop diversification and environmentally sustainable agricultural practices yields substantial advantages for the agricultural industry and the broader society (Nordhagen et al., 2021). Due to differences in both observable and unobservable characteristics of male- and female-headed families, there is a large gender gap in climate change adoption in agricultural households. It is suggested that closing the gap can boost female-headed families' climate change adoption by nearly 19% (Aryal et al., 2021).

The studies mentioned above provide evidence that gender has a role to play in the adoption of climate-smart agricultural practices in various regions of the world. The current study addresses the role gender issues play in adopting climate-smart agriculture practices and the challenges faced by the farmers in the study district of Odisha.

5.3 Methodology

The study of climate change is inherently multidisciplinary and requires collaboration across various fields within the physical and social sciences. This chapter thus uses a focus group discussion (FGD) to gain insights and understanding on the role of gender in climate change adaptation. The utilization of this tool enables the collaborative compilation and examination of data to understand the dynamics of various factors, including the implementation of a specific innovation.

Qualitative data was collected in the study area through FGDs to investigate the impact of gender on the adoption of CSA among rural farmers. The FGDs aimed to gain insight into the perspectives of both males and females regarding climate change and their inclination towards embracing climate-smart agriculture (CSA) techniques. The FGDs were conducted in 10 villages within the study area, with a participant count ranging from 12 to 15 individuals per village. The FGDs comprised household heads willing to participate and knowledgeable about the prevailing climatic conditions and associated adoption measures within their respective localities. The study's participants were 35- to 55-year-olds who possessed at least ten years of farming experience. To investigate perspectives on gender, distinct FGDs were held with male and female participants. The study was conducted during February and March of 2019. Participants were allowed to attend the study during their leisure time in the afternoon, specifically from 1 to 4 pm. Each FGD lasted for a duration of 60–90 minutes. The selection of the researcher and moderator from the same native state was made to facilitate effective communication and comprehension. Before initiating the discussion, the farmers were asked for their informed consent. The research personnel were introduced to the participants by extension agents operating within the local area. The effective communication facilitated a secure environment for the farmers to participate in the FGD and agree to be interviewed by us. The assistance of the indigenous data enumerators facilitated the discourse. The assistance of the local enumerators facilitated improved communication with the male and female members of the local community.

5.4   Livelihood and Food Security Issues

The districts of Balangir and Mayurbhanj have a large tribal population. Tribal populations rely on agriculture and forest resources for a living. The residents of these two districts make their living through agriculture and livestock. Male farmers, in general, choose agriculture as their primary source of income. The forest is a source of income for tribal farmers in Balangir and Mayurbhanj. Male farmers forage for wood, bamboo, and honey in the forest and sell them at a nearby market. Female farmers rely on the forest for their subsistence. Female farmers collect kendu leaf, jhuna, broomsticks, mahul, and other medicinal products from the forest. Some products, such as kendu leaf and mahul, are sold through cooperatives, while others are sold directly to customers in the nearest market. Traditionally, they relied on agriculture and forest resources for a living. However, slow agricultural growth and significant loss of forest cover have highlighted alternative employment structures that are insecure and unsustainable for these regions. The impediment to basic livelihood activities leads to distress migration and an increase in the region's poverty rate. The Balangir district continues to have a high inter-state and urban migration rate.

Farmers in Balangir complained that mechanization had reduced the hours they could engage as agricultural labourers. There are not enough Mahatma Gandhi National Rural Employment Guarantee Act (MGNREGA) projects in the village to keep them employed. So, they are leaving the state and migrating to find employment elsewhere. A farmer commented -

> In the Kharif season, we farm, and in the other seasons, we migrate to another place with the whole family to work in the brick industry," says one Balangir farmer. "By taking the whole family with us, we can earn more money, and after working for six months, we can return to our homeland. Therefore, we engage in seasonal migration as a means of income diversity,

Both male and female agricultural labourers participate in employment guarantee programmes such as MGNREGA. They participate in constructing village roads, ponds, and other infrastructure improvement projects within the village. A disparity in wages exists between male and female farmers in the area. A negligible disparity exists between the levels of agricultural productivity efforts exerted by males and females. The societal norms and patriarchal structure perpetuate a gender-based wage disparity for female farmers.

The inhabitants of Kendrapara district rely on marine-based livelihoods and non-agricultural employment opportunities. In the non-agricultural season, male farmers engage in fishing activities in both marine and freshwater bodies. Farmers who possessed boats exhibited greater resourcefulness compared to those who did not. However, some farmers work daily as fishing labourers and earn a restricted income. The alternate cohort of agriculturalists had procured rented boats to engage in piscatorial activities. Fishing served as a viable means of diversifying the livelihoods of the coastal farmers residing in the Kendrapara district. As reported, the Mahakalpara Block of Kendrapara district has experienced a growth spurt in the prawn industry. Farmers are employed as labourers in the prawn industry to support their livelihood. A small proportion of young members within households engage in inter-state migration, and they consistently provide financial support for agricultural pursuits within their families. Women engaged in agricultural activities in Kendrapara district tend to avoid leaving their residences for employment. Instead, they engage in domestic labour. Females engage in modest economic activities such as agroforestry, food preparation, aquaculture care and production, domestic poultry rearing, cultivation and selling of agricultural products from fields and homestead gardens, managing small-scale businesses, and labour in sectors other than agriculture.

The interviewed men and women reported food security as a major concern for their families. The climate hazards make them opt for off-farm income. Their off-farm income had to be spent on medical emergencies, fertilizers, seeds, and agricultural machine rent. However, they face food security issues due to low off-farm income and productivity during natural hazards. Due to inflation and low income, families reduce their consumption amount and change their consumption patterns. Households with low socioeconomic status experience limited availability of food resources, which increases their vulnerability to malnourishment. Child malnutrition is a prevalent health concern that increases susceptibility to mortality and morbidity related to infectious and parasitic illnesses. Access to safe water and sanitation remains fundamental for remote communities' necessities. However, the government-aided public distribution system and public health aids partially help them to recover from food insecurity.

5.5   Exposure to Climate Hazards and Impact on the Livelihood

The Indian Government's National Action Plan on Climate Change (NAPCC) acknowledges women as a particularly vulnerable group in the context of climate change. NAPCC specifies that "The impacts of climate change could prove particularly severe for women. With climate change, there would be increasing scarcity of water, reductions in yields of forest biomass, and increased risks to human health, with children, women, and the elderly in a household becoming the most vulnerable ... special attention should be paid to the aspects of gender".

The adverse impacts of climate change and natural hazards have devastated the economy of the Balangir and Kendrapara districts. The farmers in these two districts were negatively affected by the extended drought, cyclone, and flood events that the region witnessed. The FGD findings indicate that small-scale farmers of both genders have noted increased climate change-related hazards. Male and female farmers are commonly exposed to natural hazards threatening their agricultural livelihoods. The farmers in these two districts have reported experiencing significant shocks such as droughts, heat waves, typhoons, outbreaks of pests and diseases, floods, and fluctuations in crop productivity.

The women farmers of Balangir district reported that due to the drought, they failed to work in the agriculture field, which made them suffer the loss of wages which created an imbalance in their family expenditure. The women head farmer was the family's sole bread earner through agriculture, so the loss of yield due to drought made them incur financial losses and face livelihood issues. Due to drought events, women farmers had the burden of financial credit. Due to the lack of land rights, the women farmers depended on informal credit, where they were charged a heavy interest rate. So, loans from informal credit sources pushed them into more financial distress. The women farmers dependent upon the forest-based off-farm livelihood lost their way of earning. Due to drought events, forest-based products became unavailable, so the women farmers lost their livelihood. Both men and women who responded said they experienced social and health-related issues, such as disruptions in their children's education, financial difficulties, and the need to take out loans. Women described how illness outbreaks among children and household members were frequent during extreme events.

***Balangir women farmers** shared how drought and water stress in the area force them to migrate with their families. They added that it was extremely difficult for a woman to migrate to different places with small children.*

***The landless farmers of Balangir** stated that the drought affects them the most because if their crop is lost due to natural hazards, their landowners do not consider the loss. They demand money or product, so they have to migrate to compensate for the loss and repay their share to the landowners. The landowners also do not allow the tenant to claim crop insurance.*

***The farmers of Kendrapara** were also facing distress due to climate change. Farmers have discussed in FGD that the lack of water management, irrigation facilities, and drainage have been hit hard due to the agriculture practices. Rivers and tributaries flow into the district. However, due to a lack of proper planning, farmers cannot get water during the cultivation period, and instead they get flash floods during harvesting.*

A farmer from Kendrapara said, *"Pests and diseases have increased a lot in the current period; pest attacks became a threat to our yield. Government should assess the loss it and should give compensation".*

Another farmer from Mahakalpara of Kendrapara said, *"The saline water and high tide is a challenge for us; I sowed paddy seeds and transplanted saplings. The hope to get good yield was destroyed due to seawater entering our fields because of a damaged saline stone bund".*

A farmer commented on the man-made hazard, *"The increase of prawn industries in their area and the release of toxic and chemical water into the canal, combined with farmers doing irrigation, causes damage to the agricultural field and river water as well".*

A farmer in Kendrapara District's Rajnagar Block stated *sea erosion was a major issue in their area. Due to rising sea levels, the sea has washed their pasture and agricultural land away. Previously, people kept milch animals for milk. People used to keep livestock because there was pastureland available. Still, due to erosion, people have reduced their livestock and milch animal keeping, and the production of milk and dairy products has been reduced. It also posed a problem for livelihood diversification.*

According to a farmer from the coastal areas of Kendrapara District, *"the sea level is rising, extreme weather events, and changes in ocean currents have become a problem for fishing, causing a loss of livelihood".*

References

Alhassan, S. I., Kuwornu, J. K. and Osei-Asare, Y. B. (2018) 'Gender dimension of vulnerability to climate change and variability: Empirical evidence of smallholder farming households in Ghana', *International Journal of Climate Change Strategies and Management*, 11(2), pp. 195–214.

Alston, M. (2015) *Women and climate change in Bangladesh*. London: Routledge.

Alston, M. (2013) 'Women and adaptation', *Wiley Interdisciplinary Reviews: Climate Change*, 4(5), pp. 351–358.

Anderson, K., Clow, B. and Haworth-Brockman, M. (2013) 'Carriers of water: Aboriginal women's experiences, relationships, and reflections', *Journal of Cleaner Production*, 60, pp. 11–17.

Arora-Jonsson, S. (2011) 'Virtue and vulnerability: Discourses on women, gender and climate change.', *Global Environmental Change*, 21(2), pp. 744–751.

Aryal, J. P., Sapkota, T. B., Rahut, D. B., Marenya, P. and Stirling, C. M. (2021) 'Climate risks and adaptation strategies of farmers in East Africa and South Asia', *Scientific Reports*, 11(1), p. 10489.

Asadullah, M. N. and Kambhampati, U. (2021) 'Feminization of farming, food security and female empowerment', *Global Food Security*, 29, p. 100532.

Babugura, A., Mtshali, N. and Mtshali, M. (2010) *Gender and climate change: South Africa case study*. Cape Town: Heinrich Böll Stiftung Southern.

Balasha, A. M., Munyahali, W., Kulumbu, J. T., Okwe, A. N., Fyama, J. N. M., Lenge, E. K. and Tambwe, A. N. (2023) 'Understanding farmers' perception of climate change and adaptation practices in the marshlands of South kivu, democratic Republic of Congo', *Climate Risk Management*, 39, p. 100469.

Balikoowa, K., Nabanoga, G., Tumusiime, D. M. and Mbogga, M. S. (2019) 'Gender-differentiated vulnerability to climate change in Eastern Uganda', *Climate and Development*, 11(10), pp. 839–849.

Bartlett, S. (2008) 'Climate change and urban children: Impacts and implications for adaptation in low-and middle-income countries', *Environment and Urbanization*, 20(2), pp. 501–519.

Brody, A., Demetriades, J. and Esplen, E. (2008) '*Gender and climate change: Mapping the linkages,' Sussex: Institute of Development Studies.*

Bryan, E., Ringler, C., Okoba, B., Roncoli, C., Silvestri, S. and Herrero, M. (2013) 'Adapting agriculture to climate change in Kenya: Household strategies and determinants', *Journal of Environmental Management*, 114, pp. 26–35.

Carr, E. R. and Thompson, M. C. (2014) 'Gender and climate change adaptation in agrarian settings: Current thinking, new directions, and research frontiers', *Geography Compass*, 8(3), pp. 182–197.

Demetriades, J. and Esplen, E. (2010) 'The gender dimensions of poverty and climate change adaptation', *Social dimensions of climate change: Equity and vulnerability in a warming world*, 133–143.

Dankelman, I. (Ed.). (2010). Gender and climate change: An introduction. Routledge. https://www.google.co.in/books/edition/Gender_and_Climate_Change/7pr8xyafPi0 C?hl=en&gbpv=1&dq=+Dankelman,+I.+(2010)+Gender+and+climate+change:+An+in troduction.+Routledge.&pg=PR3&printsec=frontcover. URL Provided

Denton, F. (2002) 'Climate change vulnerability, impacts, and adaptation: Why does gender matter?', *Gender and Development*, 10(2), pp. 10–20.

Deressa, T. T., Hassan, R. M., Ringler, C., Alemu, T. and Yesuf, M. (2009) 'Determinants of farmers' choice of adaptation methods to climate change in the Nile basin of Ethiopia', *Global Environmental Change*, 19(2), pp. 248–255.

Edmunds, D., Sasser, J. and Wollenberg, E. K. (2013) A gender strategy for pro-poor climate change mitigation. *CCAFS Working Paper*.

FAO. (2011) 'Adapt: Framework Programme on Climate Change Adaptation', 38. Retrieved from https://shorturl.at/nBCJO

FAO. (2015) '*Climate change and food systems: Global assessments and implications for food security and trade*'. Rome. URL: http://www.fao.org/3/a-i4332e.pdf

Ge, Y., Fan, L., Li, Y., Guo, J. and Niu, H. (2023) 'Gender differences in smallholder farmers' adoption of crop diversification: Evidence from Shaanxi plain, China', *Climate Risk Management*, 39, p. 100482.

Glazebrook, T. (2011) 'Women and climate change: A case study from northeast Ghana', *Hypatia*, 26(4), pp. 762–782.

Goh, A. H. X. (2012) A literature review of the gender-differentiated impacts of climate change on women's and men's assets and well-being in developing countries. CAPRi Working Paper No. 106. Washington, DC: International Food Policy Research Institute. https://doi.org/10.2499/CAPRiWP106.

Gouse, M., Sengupta, D., Zambrano, P. and Zepeda, J. F. (2016) 'Genetically modified maize: Less drudgery for her, more maize for him? Evidence from smallholder maize farmers in South Africa', *World Development*, 83, pp. 27–38.

Hariharan, V. K., Mittal, S., Rai, M., Agarwal, T., Kalvaniya, K. C., Stirling, C. M. and Jat, M. L. (2020) 'Does climate-smart village approach influence gender equality in farming households? A case of two contrasting ecologies in India', *Climatic Change*, 158, pp. 77–90.

Hove, M. and Gweme, T. (2018) 'Women's food security and conservation farming in zaka district-Zimbabwe', *Journal of Arid Environments*, 149, pp. 18–29.

Huyer, S. (2016) 'Closing the gender gap in agriculture', *Gender, Technology and Development*, 20(2), pp. 105–116.

IPCC. (2007) 'Summary for policymakers', in Solomon, S., Qin, D., Manning, M., Chen, Z., Marquis, M., Averyt, K.B., Tignor, M. and Miller, H.L. (eds.) *Climate change 2007: The physical science basis. Contribution of working group I to the fourth assessment report of the intergovernmental panel on climate change*. Cambridge, United Kingdom and New York, NY, USA: Cambridge University Press.

Jost, C., Kyazze, F., Naab, J., Neelormi, S., Kinyangi, J., Zougmore, R. and Kristjanson, P. (2016) 'Understanding gender dimensions of agriculture and climate change in smallholder farming communities', *Clim Dev*, 8(2), pp. 133–144.

Jury, M. (2010) 'Climate and weather factors modulating river flows in southern Angola,' *International Journal of Climatology: A Journal of the Royal Meteorological Society*, 30(6), pp. 901–908.

Kakota, T., Nyariki, D., Mkwambisi, D. and Kogi-Makau, W. (2011) 'Gender vulnerability to climate variability and household food insecurity', *Clim Dev*, 3(4), pp. 298–309.

Khatri-Chhetri, A., Regmi, P. P., Chanana, N. and Aggarwal, P. K. (2020) 'The potential of climate-smart agriculture in reducing women farmers' drudgery in high climatic risk areas', *Climatic Change*, 158(1), pp. 29–42.

Lambrou, Y. and Nelson, S. (2013) 'Gender issues in climate change adaptation: Farmers' food security in Andhra Pradesh,' in *Research, Action and Policy: Addressing the Gendered Impacts of Climate Change*, pp. 189–206. Dordrecht: Springer. https://doi.org/10.1007/978-94-007-5518-5.

MacGregor, S. (2010) 'Gender and climate change': From impacts to discourses', *Journal of the Indian Ocean Region*, 6(2), pp. 223–238.

Majumder, J. and Shah, P. (2017) 'Mapping the role of women in Indian agriculture', *Annals of Anthropological Practice*, 41(2), pp. 46–54.

Mphande, E., Umar, B. B. and Kunda-Wamuwi, C. F. (2022) 'Gender and legume production in a changing climate context: Experiences from Chipata, Eastern Zambia', *Sustainability*, 14(19), p. 11901.

Murray, U., Gebremedhin, Z., Brychkova, G. and Charles, S. (2016) 'Smallholder farmers and climate smart agriculture: Technology and labor- productivity constraints amongst women smallholders in Malawi', *Gender, Technology and Development*, 20(2), pp. 117–148. https://doi.org/10.1177/0971852416640639.

Nchanji, E. B., Kabuli, H., Onyango, N., Cosmas, L., Chisale, V. and Matumba, A. (2022) 'Gender differences in climate-smart adaptation practices amongst bean-producing farmers in Malawi: The case of Linthipe extension planning area', *Frontiers in Sustainable Food Systems*, 6, p. 1001152.

Nellemann, C. Verma, R. and Hislop, L. (eds.) (2011) *Women at the frontline of climate change: Gender risks and hopes. A Rapid Response Assessment. United Nations Environment Programme and GRID-Arendal*. https://gridarendal-website-live.s3.amazonaws.com/production/documents/:s_document/165/original/rra_gender_screen.pdf?1484143050

Nnadi, O. I., Lyimo, J., Liwenga, E. and Madukwe, M. C. (2023) 'Gender perspectives of responses to climate variability and change among farm households in Southeast Nigeria,' *Environmental Management*, 71(1), pp. 201–213.

Nordhagen, S., Pascual, U. and Drucker, A. G. (2021) 'Gendered differences in crop diversity choices: A case study from Papua New Guinea', *World Development*, 137, p. 105134.

Ogunlela, Y. I. and Mukhtar, A. A. (2009) 'Gender issues in agriculture and rural development in Nigeria: The role of women', *Humanity and Social Sciences Journal*, 4(1), pp. 19–30.

Omolo, N. A. (2010) 'Gender and climate change-induced conflict in pastoral communities: Case study of Turkana in northwestern Kenya', *African Journal on Conflict Resolution*, 10(2). https://doi.org/10.4314/ajcr.v10i2.63312

Onta, N. and Resurreccion, B. P. (2011) 'The role of gender and caste in climate adaptation strategies in Nepal', *Mountain Research and Development*, 31(4), pp. 351–356.

Peterman, A., Behrman, J. A., & Quisumbing, A. R. (2014). A review of empirical evidence on gender differences in nonland agricultural inputs, technology, and services in developing countries (pp. 145–186). Springer Netherlands.

Quisumbing, A. R., Brown, L. R., Feldstein, H. S., Haddad, L. and Peña, C. (1996) 'Women: The key to food security', *Food and Nutrition Bulletin*, 17(1), pp. 1–2.

Ribeiro, N. and Chaúque, A. (2010) *Gender and climate change: Mozambique case study*. Cape Town: Heinrich Böll Stiftung Southern Africa.

Rola-Rubzen, M. F., Paris, T., Hawkins, J. and Sapkota, B. (2020) 'Improving gender participation in agricultural technology adoption in Asia: From rhetoric to practical action', *Applied Economic Perspectives and Policy*, 42(1), pp. 113–125.

Steinfield, L. and Holt, D. (2020) 'Structures, systems and differences that matter: Casting an ecological-intersectionality perspective on female subsistence farmers' experiences of the climate crisis', *Journal of Macromarketing*, 40(4), pp. 563–582.

Sugden, F., Maskey, N., Clement, F., Ramesh, V., Philip, A. and Rai, A. (2014) 'Agrarian stress and climate change in the Eastern Gangetic plains: Gendered vulnerability in a stratified social formation', *Global Environmental Change*, 29, pp. 258–269.

Tall, A., Hansen, J., Jay, A., Campbell, B. M., Kinyangi, J., Aggarwal, P. K. and Zougmoré, R. B. (2014) Scaling up climate services for farmers: Mission possible. Learning from good practice in Africa and South Asia. *CCAFS Report*.

Tantoh, H. B., McKay, T. T., Donkor, F. E. and Simatele, M. D. (2021) 'Gender roles, implications for water, land, and food security in a changing climate: A systematic review', *Frontiers in Sustainable Food Systems*, 5, p. 707835.

Theis, S., Sultana, N. and Krupnik, T. J. (2018) *Overcoming gender gaps in rural mechanization: Lessons from reaper-harvester service provision in Bangladesh (CSISA PROJECT No. 9)*. International Food Policy Research Institute (IFPRI).

Tschakert, P. and Machado, M. (2012) 'Gender justice and rights in climate change adaptation: Opportunities and pitfalls', *Ethics and Social Welfare*, 6(3), pp. 275–289.

UN-Women (2015) *Women in power and decision-making. The Beijing platform for action turns 20*. New York: United Nations.

Vemireddy, V. and Choudhary, A. (2021) 'A systematic review of labor-saving technologies: Implications for women in agriculture', *Global Food Security*, 29, p. 100541.

Yadav, S. S. and Lal, R. (2018) 'Vulnerability of women to climate change in arid and semi-arid regions: The case of India and South Asia', *Journal of Arid Environments*, 149, pp. 4–17.

Ylipaa, J., Gabrielsson, S. and Jerneck, A. (2019) 'Climate change adaptation and gender inequality: Insights from rural Vietnam', *Sustainability*, 11(10), p. 2805.

6 The Impact of CSA Technology Adoption on Natural Capital

6.1 Introduction

Natural capital and climate-smart agriculture (CSA) are profoundly interconnected, impacting how we think about environmental preservation and sustainable food production. CSA refers to practices and techniques that aim to increase agricultural productivity, enhance resilience to climate change, and reduce greenhouse gas (GHG) emissions. On the other hand, natural capital encompasses the Earth's ecosystems, such as forests, wetlands, and biodiversity, which provide vital services to support human well-being.

Natural capital and CSA interact because they both aim to promote sustainable development. Agroforestry, precision farming, and water management are climate-smart agricultural practices intended to maximize resource utilization, reduce environmental impact, and increase output. These procedures actively utilize and safeguard natural resources by preserving soil fertility and water quality.

Global population growth, urbanization, and unsustainable human activities have led to rapid environmental changes and alterations in climatic conditions worldwide. This development and progress come at the expense of natural capital, which is limited and continuously depleting, posing a significant global challenge. Therefore, it is crucial to strike a balance between the development of humanity and the preservation of natural capital. Agriculture, a significant contributor to climate change and natural capital imbalance, can cause and be affected by these issues. Climate change can impact food systems through various mechanisms, including direct effects on crop production (such as droughts, floods, and growing season length), market dynamics, fluctuations in food prices, and modifications in supply chain infrastructure.

In developing nations like India, agriculture is essential to guaranteeing food security, nutrition, and means of subsistence. Agriculture is the main source of industrial raw materials, supporting two-thirds of the population and considerably contributing to the nation's gross domestic product. Conventional agricultural methods, on the other hand, have resulted in decreased factor productivity, degradation of natural resources, deterioration of soil health,

DOI: 10.4324/9781003290445-6

shortage of water, groundwater depletion, and increased susceptibility to diseases and pests (Bhatt et al., 2016; Meena and Meena, 2017).

Despite the transformative impact of Green Revolution technologies in India, food insecurity, malnutrition, poverty, and hunger persist. According to the 2017 Global Hunger Index, India ranks 100th out of 119 countries, with a score of 31.4, denoting a "severe" degree of hunger (International Food Policy Research Institute (IFPRI, 2017). CSA is used to manage climate risks and guarantee consistency and sustainability in agricultural production. Adopting CSA techniques and technology, according to studies conducted in Ethiopia, increases food security (Shiferaw et al., 2014).

Sustainable agriculture encompasses various practices that utilize land and water resources to meet current food demands while considering future generations. Its primary objectives are to develop environmentally beneficial and economically viable agroecosystems that sustain long-term arable land productivity. Sustainable agriculture involves profitable, efficient, and productive farming practices that align with socioeconomic factors. Key pillars of sustainable agriculture include enhanced nutrient cycling, improved soil quality, reduced land degradation, integrated nutrient and pest management (INPM), and efficient marketing strategies to enhance profitability. World is currently facing two critical challenges: climate change and biodiversity loss. Addressing these challenges requires significant political and socioeconomic transformations at individual and institutional levels to transition to a low-carbon economy and reduce GHG emissions in line with the obligations of the Paris Agreement. Additionally, there is a pressing need to adapt to unavoidable climate change impacts. The increasing global demand for food, driven by changing diets and a growing population, exacerbates the strain on production as food yields plateau, ocean quality declines, and natural resources such as soils, water, and biodiversity are depleted.

According to projections from 2020, there are now almost 690 million hungry people worldwide, or 8.9% of the world's population, an increase of around 60 million over the previous five years. The food security problem will only worsen as there will be a projected need to produce approximately 70% more food by 2050 to feed an anticipated 9 billion people. Agriculture is particularly vulnerable to the impacts of climate change. Rising temperatures, unpredictable weather patterns, shifting agroecosystems, invasive species, and extreme weather events already negatively affect animal productivity, nutritional value of staple crops, and crop yields. Significant investments in adaptation will be necessary to maintain current yields, enhance productivity, and ensure food quality to meet the growing demand.

However, agriculture contributes significantly to the climate crisis, accounting for 19–29% of global GHG emissions. Without action, the percentage could increase significantly as emissions decrease in other industries. Additionally, one-third of the food produced worldwide is wasted, highlighting the importance of addressing food loss and waste to achieve climate targets and reduce environmental stress. Approximately 38% of the world's land is used in agriculture, with one-third dedicated to crop production and the remainder for livestock grazing. As agriculture and forestry contribute to 23% of global GHG emissions, it is crucial to implement changes in land management practices while safeguarding food production and farmers' livelihoods. Considering the increasing challenges to food security due to strained natural resources and climate change, it is essential to reevaluate land management and food production approaches.

The current production systems pose a significant threat to sustainable agriculture. Food systems worldwide release greenhouse gases like carbon dioxide (CO_2), nitrous oxide (N_2O), and methane (CH_4), contributing to approximately one-third of global warming. The issue is exacerbated by unsustainable water consumption, mainly driven by agriculture as the largest consumer. Additionally, conventional farming practices have caused soil degradation, habitat devastation, and a decline in biodiversity. Consequently, the foundations of current production methods are deteriorating, posing a threat to food security and resource management.

In summary, existing agricultural production technologies need to be improved to ensure long-term food security and sustainable natural resource management. It is imperative to adopt transformative measures to mitigate climate change impacts, promote sustainable land and water management practices, and transition towards resilient and environmentally friendly agricultural systems.

Globally, there is a growing trend of increasing economic losses caused by natural disasters, with the agriculture industry particularly vulnerable to them. The "United Nations Office for Disaster Risk Reduction" (UNISDR, 2018) reported that countries affected by disasters incurred direct economic losses amounting to US$2908 billion between 1998 and 2017. A significant proportion of these losses, 77%, were attributed to climate-related disasters. The agriculture sector has experienced notable impacts from climate change in recent years.

In India, the anticipated effects of climate change on major crop yields are a significant issue that requires attention, with projections indicating a potential countrywide decline of up to 9% between 2010 and 2039. This decline is expected to exacerbate over time, posing substantial challenges to agricultural productivity. Specifically, the potential yield losses vary across crops and locations, depending on the geographic region and projected climatic conditions, from 35% for rice, 20% for wheat, 50% for sorghum, and 13% for barley, to as much as 60% for maize.

- In the context of Indian agriculture, which heavily relies on monsoons, the productivity of crops remains highly vulnerable to the influence of climate, particularly the fluctuations in temperature and rainfall patterns. These climatic variations pose a significant threat to food security in India.
- In recent decades, India has experienced a discernible impact of climate change, leading to a temperature increase ranging from 0.6°C to 25.1°C between 1901 and 2018. These changes have also induced shifts in monsoon patterns, further complicating agricultural conditions.

- Projections suggest that by the year 2100, the productivity of most crops in India is likely to decline by 10–40% due to the combination of rising temperatures, unpredictable rainfall, and reduced availability of irrigation water.
- In India, about 60% of cropland is used for rainfed or unirrigated crops, which are at high risk of being negatively impacted by climate change. A slight increase of 0.5°C in winter temperature is projected to cause a 0.45-ton reduction per hectare in rainfed wheat yield.
- According to the Government of India's economic survey in 2018, the negative impacts of climate change have led to an annual economic loss of $9–10 billion in the country.

6.2   Natural Capital and Its Interaction with Agriculture and Climate Change

6.2.1   *Natural Capital: Components and Importance*

Natural capital is a concept that emphasizes the intrinsic value of the Earth's ecosystems and the services that they offer to support human well-being. It encompasses various components that contribute to the functioning and resilience of ecosystems. These components include:

1 Biodiversity: Biodiversity represents the variety of living organisms, including plants, animals, and microorganisms, within an ecosystem. It plays a crucial role in maintaining ecosystem balance, providing essential services such as pollination, pest control, and nutrient cycling.
2 Ecosystems: Ecosystems are dynamic living systems that encompass diverse environments like forests, wetlands, grasslands, and oceans, which undergo continuous changes in environments, each with its unique characteristics and functions.
3 Soil is a fundamental component of natural capital, serving as the foundation for terrestrial ecosystems. It supports plant growth, nutrient cycling, water filtration, and carbon storage.
4 Water: Freshwater ecosystems, including rivers, lakes, and wetlands, are vital components of natural capital. They provide clean water for human consumption, support aquatic biodiversity, regulate water flow, and act as natural buffers against floods and droughts.

Natural capital is of great importance due to the essential services it offers to society. These services encompass the supply of valuable resources like food, timber, and medicinal plants and the management of diverse ecosystem functions.

6.2.2   *Role of Natural Capital in Sustainable Agriculture in India*

The role of natural capital in sustainable agriculture in India is paramount for ensuring long-term agricultural productivity, environmental sustainability, and resilience to climate change. Natural capital refers to the Earth's ecosystems, including forests, wetlands, and biodiversity, which provide essential services such as nutrient cycling, soil fertility, pollination, water regulation, and pest control.

Agriculture in India heavily depends on ecosystem services provided by natural capital. For instance, pollinators play a crucial role in crop production, contributing to the yield and quality of various crops, such as fruits, vegetables, and oilseeds (Rai et al., 2015). Furthermore, forests and wetlands act as carbon sinks, mitigating climate change by sequestering atmospheric CO_2 (Dadhwal et al., 2014).

The conservation and restoration of natural capital are vital for sustainable agriculture in India. Protecting and managing forests, wetlands, and biodiversity contribute to preserving ecosystem services and providing habitats for beneficial organisms, enhancing soil fertility, regulating water availability, and supporting natural pest control. By recognizing the value of natural capital and integrating its conservation into agricultural practices, India can promote sustainable and resilient farming systems that ensure food security, environmental sustainability, and the well-being of rural communities.

Since 1750, Concentrations of greenhouse gases, such as nitrous oxide, carbon dioxide, and CH_4, have dramatically grown. The main cause of GHG emissions, carbon dioxide, increased from 22.15 billion metric tonnes in 1990 to 36.14 billion in 2014 (15 May 2020; NASA Earth Observatory).

Since 1975, the average global temperature has been increasing at a rate of 0.15–0.20°C each decade, and by 2021, that pace is expected to rise to 1.45–5.8°C. With fossil fuels accounting for 65% of GHG emissions, CO_2 is a significant contributor (Arora et al., 2005)

6.2.2.1 Land Use Change and Soil Degradation

One of the primary factors leading to the conversion of forested areas into croplands and pastures is land degradation. This process involves farmers depleting available land resources and looking for other, better places to plant their crops. Due to the limited ability to expand pastures into less good locations, areas with agricultural potential are changed over to grazing lands (Asner et al., 2004). Various processes, including soil erosion, volatilization, and agricultural waste burning, can deplete soil carbon. Furthermore, changes in agricultural land use and management techniques also reduce soil carbon. These factors combined result in soil carbon depletion, negatively impacting soil health, fertility, and the long-term potential for carbon sequestration (Bolin et al., 1983).

In arid regions with dry soil conditions, soil erosion becomes a significant factor in land degradation. It results in less effective infiltration, water retention, and surface runoff. The land is considerably degraded due to these combined consequences, which also harm the ecosystems and agriculture in these regions (Stroosnijder, 2007)

6.2.2.2 Water Scarcity and Impacts on Agriculture

According to a worldwide study of water scarcity, up to two-thirds of the world's population might face a water shortage in the upcoming decades (Wallace and Gregory, 2002). Humphreys et al. (2010), discuss the factors leading to water depletion in the rice-wheat belt of North-West India. It explains that when water is used for irrigation, a portion is lost to the atmosphere through plant transpiration and evaporation from the soil surface. The remaining water either drains to the surface or subsurface storage, depending on the condition of the water. Water depletion occurs when water flows to sinks and cannot be recovered. The water table decreased from 0.2 m/year between 1973 and 2001 to about 1.0 m/year from 2000 to 2006, particularly in central Punjab. By 2025, the amount of water per person in India would have decreased to 1000 m^3, and no necessary action would have been taken (UNEP, 2008). Around 5.6 million acres in South Asia were using various resource conservation technologies (RCTs), "National Agricultural Research and Extension Systems". Promoting CA is also intended to increase the effectiveness of nutrient and water utilization (WUE). With only a $3.5 million investment in zero-tillage (ZT) technology and a 66% internal rate of return, India can save USD 164 million overall.

Water availability is crucial for crop production; factors like rainfall patterns, soil quality, and water management impact agricultural productivity. Climate change introduces uncertainties and interactions between water and agriculture, requiring adaptation programmes.

Droughts, worsened by global warming, are increasing in frequency and intensity in regions like Africa, Southern Europe, the Middle East, the Americas, Australia, and Southeast Asia (Fraser et al., 2013). Rising water demand, population growth, urban expansion, and conservation efforts exacerbate their impact. Droughts cause crop failures, loss of grazing land, and severe famines due to long-term water scarcity. Heatwaves further harm crops through heat stress and wilting. Agricultural land affected by drought has risen for significant crops (Rosenzweig, 2007).

Livestock grazing suffers from reduced water and grass availability, increasing feed demand and prices. Heavy rainfall and flooding damage crops, delay farming and reduce grain quality. Rising CO_2 levels amplify rainfall intensity. Flooding harms vegetable crops and higher temperatures worsen symptoms. Tropical storms impact countries like the USA, China, and India and cause crop destruction, land loss, and displacement. Extreme weather changes may harm crop growth, highlighting agricultural vulnerability to climate change.

6.2.2.3 *GHGs and Their Impacts on Agriculture*

Tillage, land usage, fertilizer, manure application, crop burning, and other soil management methods contribute to CO_2 generation in agriculture. These techniques cause the breakdown of organic materials in the soil and the emission of CO_2 gas. The wetland environment is the main contributor to CH_4 emissions among the various agricultural habitats.

Using biochar in agriculture opened the door for its long-term, undistributed storage in the soil, and scientists have predicted that it will lower GHG emissions. A very caseous by-product of biomass pyrolysis is biochar. Purakayastha et al. (2015) found that applying cornstalk biochar significantly reduced the CH_4 emission from rice soil.

Further, Jalota et al. (2007) identified 7–40 cm of irrigation water savings with mulching. The advantages of mulch depending on the season, irrigation schedule, consistency of the soil, and type of mulching material. It was discovered that (Jalota and Arora, 2002; Dadhich et al., 2015) crop variety is crucial for reducing the quantity of irrigation water needed. The system as a whole saw an improvement in water production, especially when rice was diverted.

Global warming is being caused by greenhouse gases, which have further impacted agricultural productivity in some way. New technologies, such as slow-release fertilizers, direct-seed rice, and soil test-based fertilization, will undoubtedly offer the potential to lower GHG emissions. CO_2, CH_4, and N_2O overproduction are the major causes of global warming. CH_4 makes up 18% of all GHGs. The primary areas on which we must concentrate to reduce the production of these GHGs to minimize the harm caused by global warming are crop management, animal management, bioenergy, grassland management, manure/biosolid preparation, and land management (Smith et al., 2008).

India intends to cut its GHG emission intensity by 20% by 2020 per the "United Nations Framework Convention on Climate Change" (UNFCCC) policy. Global climate change results from the current changes in increasing sea levels and glacier melting. The leading causes of GHG emissions from agriculture in a growing nation like India are excessive fertilizer use, other agricultural inputs, and an expanding animal population.

6.2.3 Cost of Natural Capital

The world has emitted approximately 1.5 trillion metric tons of CO_2 since 1751, with Europe being the largest contributor. Climate change has positive and negative impacts, including increased crop fertilization and decreased energy requirements, but it also negatively affects water resources. In the 21st century, climate change is expected to become a severe problem affecting developed and developing countries (Figure 6.1) (Tol, 2013).

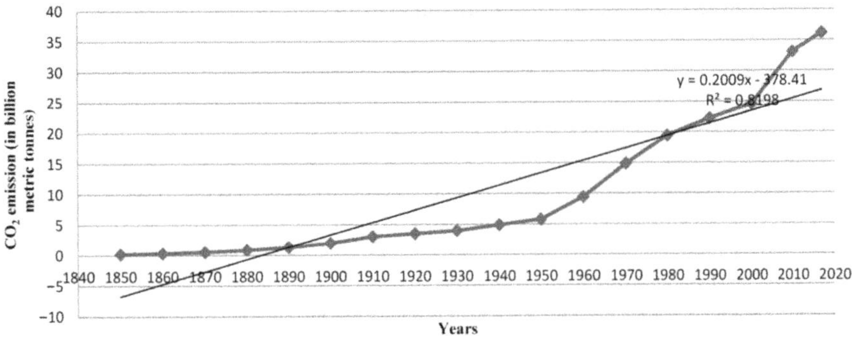

Figure 6.1 CO_2 Emission in the Atmosphere Over the Years (1850–2020.)
Source: Our World in Data.

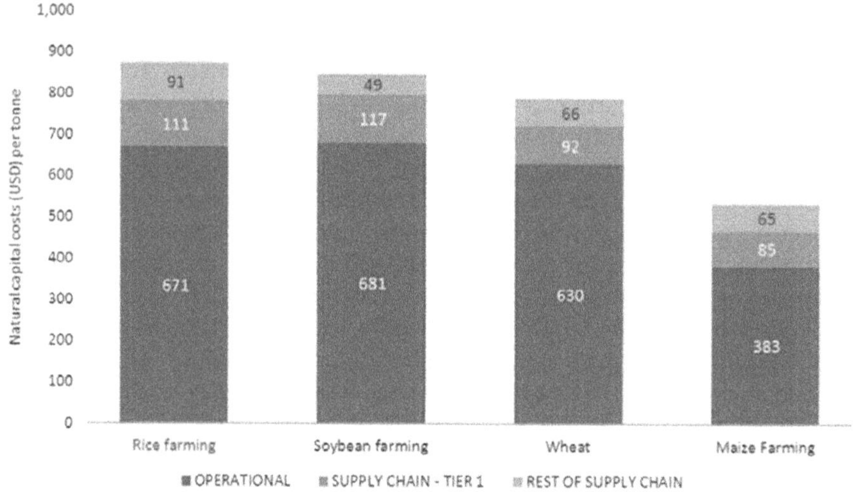

Figure 6.2 Global Operational and Supply Chain Natural Capital Costs per Tonne of Crop Production.

Figure 6.2 demonstrates that rice production carries the highest natural capital costs among all analysed crops like soybean and wheat on a global scale; the primary factor driving these costs is the operational impacts that occur at the farm level, accounting for 77% of the total impacts. This understanding allows businesses to identify where each type of crop production generates the most significant effects along the value chain.

Table 6.1 shows that rice production contributes to the highest natural capital cost. Across all agricultural sectors, operational natural capital costs contribute 77% to the consequences. According to this study, rice cultivation has the most significant effects on natural capital in China and India. Each nation is in charge of 22% of the costs to the environment incurred by the

Table 6.1 Total Natural Capital Cost Country-Wise

Crop	Top	Contributors to the natural	Capital costs (USD million)	Total costs (USD million)
	1st	*2nd*	*3rd*	
Maize	China 129,607 (−35%)	USA 89,687 (24%)	Brazil 57,344 (16%)	369,243
Rice	China 113,681 (22%)	India 113,315 (22%)	Indonesia 62,580 (12%)	507,308
Soybean	Brazil 102,268 (58%)	USA 52,051 (30%)	Argentina 21,724 (12%)	176,044
Wheat	China 84,602 (19%)	India 69,387 (16%)	Germany 61,560 (14%)	439,188

Source: FAO (2015)

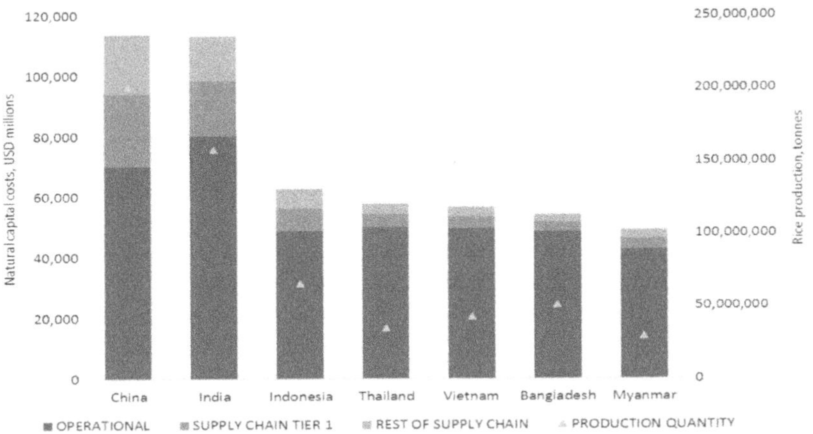

Figure 6.3 Operational and Supply Chain Natural Capital Costs of Rice Production by Country.

production of rice. Land use change, water use, and water pollution emissions all contribute to the cost of natural capital effects in the studied nations. Over 93% of the consequences in each country are attributable to these causes.

Moreover, from Figures 6.3 and 6.4, China, India, and Germany are among the top contributors to natural capital costs. The two countries that produce the most wheat, China and India, are anticipated to have a more significant influence on natural capital. Germany has a disproportionately high effect on natural capital despite producing the ninth most wheat worldwide.

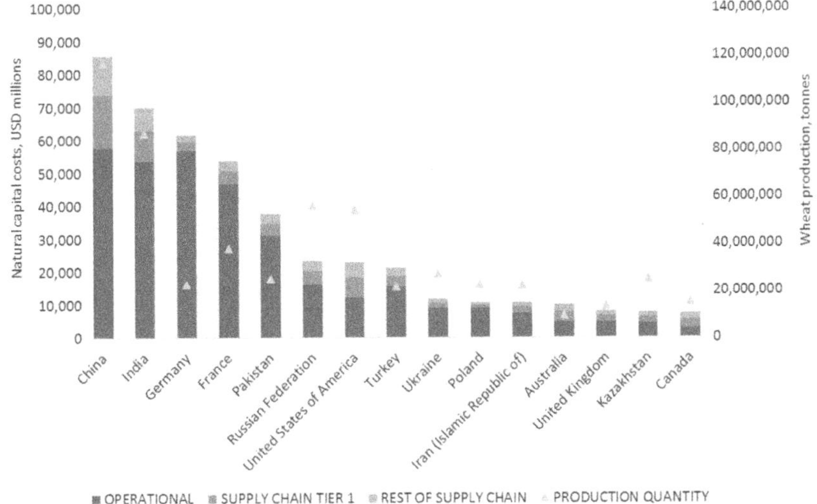

Figure 6.4 Operational and Supply Chain Natural Capital Costs of Wheat Production by Country.

6.3 Integration of Climate-Smart Agriculture and Natural Capital

6.3.1 CSA Practices for Natural Capital Conservation

CSA practices combine sustainable agricultural methods with climate change adaptation and mitigation strategies to improve ecosystem services, protect biodiversity, and foster resilient agricultural systems. This study conducts an extensive literature review to explore the primary CSA practices for conserving natural capital and assess their advantages. Additionally, it identifies the obstacles and prospects involved in implementing these practices across various agricultural systems. The findings of this review underscore the potential of CSA practices in advancing sustainable land management and biodiversity conservation while effectively addressing the challenges posed by climate change.

6.3.2 Conservation Agriculture and Soil Health Management

Natural resources are continuously depleted due to intensive farming methods that return less organic matter (OM) to the soil and continual mono-cropping systems. Along with OM depletion, erosion and soil salinization greatly exacerbate the issues. The difficulties are also worsening because of climate change's adverse effects, increased input costs, and unstable market pricing for agricultural commodities. Conservation agriculture (CA) has the potential to be a beneficial approach that uses crop variety to increase agricultural output while minimizing negative environmental consequences, minimizing soil disturbance, and permanently covering the soil with crop leftovers or protecting crops (Jena, 2019). Intensive agriculture is associated with high input levels and higher predicted yield per unit of agricultural area. High cropping intensity; a short fallow season; increased use of fertilizer, irrigation, and labour; and higher crop yields per unit of land area are typical features of intensive agriculture. Because of the limited agricultural land available, the subsistence farming system is under strain due to the rapid growth in population and concomitant high food demand (Ram and Meena, 2014).

Integration of various methods, either indirectly (soil test-based fertilization, integrated nutrient management, leaf colour chart, and slow-release fertilizers (neem-coated urea or poly-coated urea)) or directly (laser levelling, short- or medium-duration crop cultivars, transplantation/sowing time, 2-day intermittent irrigation, soil matric potential-based irrigation, permanent beds, rice without puddling (DSR), transplanting of rice with machines, etc., An method that depends on the diverse soil and climatic circumstances are to increase the declining land and water production in the area (Bhatt and Kukal, 2014; Verma et al., 2015).

Crop variety, when correctly integrated helps to reduce the quantity of irrigation water needed to minimize the effects of global warming (Dadhich et al., 2015). Crop rotation has a less significant impact on soil carbon compared to tillage techniques. Increasing the output of carbon and biomass from various crops grown impacts the soil carbon by diversifying the root pattern and depth.

Controlling the ambient temperature is mainly impractical in systems with heavy grazing. It is more effective when the animals are provided with ample shade, drink, and a place to wallow. When shade trees are around, animals are less affected by the heat and work more. Additionally, trees improve the amount and quality of pasture, which can reduce overgrazing and halt soil erosion (Thornton and Herrero, 2010). Initiatives including cross-breeding can aid in resiliency, food security, and mitigation. In contrast to pure shorthorn breeds that had historically dominated these difficult environments, composite cattle breeds that were recently produced in tropical grasslands of northern Australia have demonstrated superior heat tolerance and disease resistance, as well as better fitness and reproductive qualities (Bentley et al., 2008)

6.3.3   *Sustainable Livelihood*

The conventional responsibilities played by researchers and crop farmers must be significantly different for sustainable livelihood (SL) implementation to be successful. Experts can aid in the empowerment of farmers by setting up farmer meetings, facilitating information about technological options, equipment, seeds, and planting options, as well as integrated pest management, soil conservation, and the systematic application of traditional knowledge to mitigate the effects of climate change (Chambers, 1991).

In July 2004, the Mizoram legislative assembly in India enacted the Organic Farming Bill, marking a significant legislative milestone in promoting and regulating organic farming practices within the region. This legislative measure aimed to support and encourage the adoption of organic farming methods while providing a framework for oversight and governance of organic agriculture in Mizoram. The passing of this bill reflects the government's commitment to promoting sustainable agricultural practices and recognizing the importance of organic farming for environmental conservation, human health, and socioeconomic development in the region. To ensure sustainable soil fertility and minimize reliance on external agricultural chemicals such as fertilizers and pesticides, agricultural practices incorporate the utilization of crop residues, organic manure, bio-fertilizers, and biological pest control methods. These approaches aim to enhance soil health and promote environmentally friendly farming practices ("Agriculture Department of Mizoram").

6.3.4   *Biodiversity*

Living beings respond to their surroundings differently, adapting, avoiding danger, and using survival techniques. They regulate their body temperature through heating, cooling, hibernation, migration, and modifying physical attributes like growing hair or changing pigments. Consequently, the ideal environmental temperature differs across organisms and crops. Climate change's diverse elements, including temperature, CO_2 levels, humidity, heat waves, and water availability, are expected to impact biodiversity at different levels, from individual organisms to entire biomes. Climate change may alter the length of a plant's vegetative, reproductive, or blooming stage and the number of available insect pollinators; this mismatch might lead to the extinction of both species (Rafferty and Ives, 2010).

According to observations, the timing of seasonal migrations, fruiting, and blooming has been vulnerable to changes in response to climatic conditions, notably temperature and rainfall. These phenological changes in various organisms highlight the influence of climate variations on their life cycle and behavioural patterns (Parmesan, 2006).

In India, rainfed agriculture covers around 70 million hectares of cultivable land in regions including Madhya Pradesh, Uttar Pradesh, Maharashtra, Karnataka, Andhra Pradesh, and Tamil Nadu. To achieve sustainability, an integrated approach involving rainwater harvesting and drainage systems can effectively manage soil and crops. In order to adapt to climate change, it is essential to develop a robust variety of sorghum, pearl millet, maize, cassava, and dual-purpose grain legumes like cowpea. To combat population pressure, agriculture must quickly embrace resource conservation practises, soil erosion, and climate uncertainties. Sustainable crop production intensification presents opportunities to optimize productivity while considering social, political, environmental, and economic factors. The mitigation of GHG emissions, advancement of adaptation science, and their implementation will significantly influence human endeavours in the twenty-first century.

6.3.5 *Water Use Efficiency and Watershed Management*

Water scarcity will reduce agricultural production by irregular rainfall, length, and distribution at local, regional, and global levels, lowering agricultural productivity (Antle and Capalbo, 2010; Falloon and Betts, 2010). Under climate change scenarios, droughts can have multiple negative impacts. These include reducing pasture productivity, limiting irrigation possibilities, increasing soil erosion and degradation through wind erosion, and potentially leading to livestock deaths due to limited water and forage availability (Gomez, 2005).

It is critical to use water from the underground effectively for restricted irrigation. Through mixed/inter-cropping, more rainfall during the Kharif season may be used. In dryland farming, ridge planting increases plant population and productivity. After the summer has passed, winter crops can be planted. Moisture is preserved in loamy sand soils via straw mulching. Crop residue management increases soil organic matter, lowers runoff, erosion, and nutrient losses, reduces the demand for irrigation, and changes albedo and GHG emissions (Post et al., 2000; Ric Powlson et al., 2001; Berndes et al., 2003; Falloon et al., 2004; Huntington, 2006). The elevation of sea levels can result in the degradation of agricultural land, particularly in areas characterized by lower altitudes or prone to flooding (IPCC, 2007) and an increase in groundwater and soil salinity as a consequence (Motha, 2007).

6.4 Some Effective CSA Techniques and Their Impact

In this section, we will explore the impact of various effective CSA techniques. These techniques play a crucial role in enhancing agricultural productivity, reducing emissions, and improving resilience in the face of climate change.

6.4.1 Crop Intensification

The green revolution's focus on high-yielding varieties, particularly rice and wheat, these cereals gained popularity due to their ease of cooking, taste, and soft texture, especially among older individuals and children. Soybean, sunflower, cotton, and cash crops benefited from favourable policies, increased consumer demand, high prices, and high yields. However, promoting oilseeds and pulses cultivation reduced farmers' interest in growing coarse cereals, as their storage capacity in rural areas needed to be improved.

6.4.2 Managing Crop Residues for Soil Carbon Sequestration and Conservation

Research indicates that soil productivity can be improved by increasing soil organic carbon (SOC) levels through direct seeding and sustainable management of crop residues. These approaches enhance soil fertility, nutrient availability, and overall productivity by incorporating OM into the soil (Lal, 2005). Enhancing SOC levels improves soil's physical structure and chemical properties, leading to enhanced nutrient availability for plants. CO_2 is captured by green plants through photosynthesis and converted into sugars, hemicellulose, cellulose, and lignin. This carbon is then cycled through the biosphere, reaching the soil as dead plants, root exudates, and deceased animals. Soil microbes decompose this organic matter, releasing CO_2 into the atmosphere through respiration.

6.4.3 Effective Crop Rotations

Crop rotation is an economically viable approach to achieving food security while minimizing environmental impacts. The selection of crop species should be guided by optimizing soil resource utilization, considering its pivotal role as a limiting factor in crop production. Implementing crop rotation methods, characterized by extended crop cover periods and reduced tillage intensity, can contribute to the enrichment of SOC content.

A critical development that can result in better crop residue management, increase SOC levels, and mitigate climate change is the use of forages in crop rotations. The increasing nitrogen levels in the soil, enabled by nitrogen-fixing legumes (80–135 kg N ha^{-1}), or the decreased nitrogen removal brought on by sunflower agriculture can be blamed for the improved sorghum yields reported (Varvel and Wilhelm, 2003; Kaye et al., 2007)

Effectively implementing crop rotation management and accurately applying fertilizers are essential for reducing residual soil nitrate levels and mitigating N_2O emissions. Moreover, farmers can actively influence biological nitrogen fixation by introducing Rhizobia strains with nitrogen-fixing and phosphorous-solubilizing abilities to legume crops (Kumari, 2017).

6.5 Case Studies: Implementation of CSA and Its Benefits

Table 6.2 provides a comprehensive overview of climate-smart technologies and their economic benefits in agricultural systems across different regions. These case studies shed light on the potential of adopting innovative approaches to enhance productivity, conserve resources, and improve economic outcomes in farming.

The first set of case studies focuses on rice cultivation in Vietnam, the Philippines, and India. Implementing site-specific nutrient management techniques resulted in increased partial factor productivity of nitrogen, leading to improved yields. The associated economic benefits varied across the regions, with increments of 34 US$/ha in Vietnam, 106 US$/ha in the Philippines, and 168 US$/ha in India. This highlights the effectiveness of tailored nutrient management strategies in optimizing crop production and profitability.

Moving on to wheat production in Sindh, Pakistan, the adoption of laser land levelling technology demonstrated significant water savings and reduced irrigation time. This not only addresses water scarcity issues but also reduces input costs for farmers, leading to an economic benefit of INR 23,250 per acre. The case study emphasizes the importance of precision land management practices in conserving water resources and improving economic returns.

In Punjab, Pakistan, the focus shifts to the rice-wheat cropping system. The implementation of zero tillage and bed furrows resulted in higher water productivity, reduced irrigation water usage, and improved fertilizer efficiency. Although specific economic benefits are not provided, the adoption of these climate-smart technologies shows promising potential in terms of resource conservation and productivity enhancement.

The Nyando basin case study in Kenya offers a broader perspective by incorporating multiple crops and livestock. The use of stress-tolerant crop

Table 6.2 Case Studies of CSA Technology Adaption, Its Benefits and Efficiency

Location	Crop	Climate-smart Technology	Enhanced efficiency	Incremental Economic Benefit	References
Vietnam Philippines India	Rice	Site-specific nutrient management	Increased partial factor productivity of nitrogen	34 US$/ha 10 6US$/ha 168 USS/ha	Pampolino et al. (2007)
Sindh, Pakistan	Wheat	Laser land levelling	Saving of 21% irrigation water and reduced irrigation time	INR 23,250/acre	Wagan et al. (2015)
Punjab, Pakistan	Rice-wheat cropping system	Zero tillage Bed furrows	Higher water productivity, saving of irrigation water, and higher fertilizer use efficiency		Latif et al. (2013)
Nypndp basin of Kenya	Multiple crops and livestock	Stress-tolerant crop varieties Improved	Laser land levelling Increased household income leading to household asset accumulation and investment livestock breeds	Increased HH income by 83% Increased HH income by 76%	Ogada et al. (2020)
Semi-arid tropics of India	Groundnut	Drought-tolerant varieties	Increase in yield by 23%, lower variability in yield, increased share of risk benefits in total benefits	17% reduction in variable cost	Birthal et al. (2012)
Kamal, Haryana	Wheat	Zero tillage	Enhanced production by 1.88% and lower cultivation cost	Higher net income	Tripathi et al. (2013)
North-western Indo-Gangetic plains of India	Rice and wheat	Laser land levelling	Reduced irrigation time, increased yield, reduced electricity charges	USS 143.5/ha/year	Aryal et al. (2015a)

(Continued)

Table 6.2 (Continued)

Location	Crop	Climate-smart Technology	Enhanced efficiency	Incremental Economic Benefit	References
North western India	Wheat	Zero tillage	Reduced cultivation cost, reduced GHGs emissions, and increased yield	USS 97.5/ha	Aryal et al. (2015b)
Upper Gangetic plains	Wheat	Site-specific nutrient management	Increased yield by 29% over farmers' fertilizer practices (FFP)	INR 68,980/ha over FFP	Singh et al. (2015)
Indo-Gangetic plains of India	Rice-Wheat cropping system	Improved crop varieties Zero- tillage	Increased net returns Laser land levelling	INR 15,712/ha/yj INR 8,119/ha/jx INR 6,951/ha/yx	Khatri-Chhetri et al. (2016)
India	Rice	Direct-seeded rice	Reduced irrigation and preparation costs	Increase HH income by 16%	Mishra et al. (2016)
Tamilnadu, India	Okra	Drip irrigation	Saving of irrigation water and electricity charges, reduced cultivation cost	INR 5,050-INR 8, 100/ha over	Narayanamoorthy and Devika (2017)
Punjab, India	DSR-Wheat	Direct-seeded rice	Saving of irrigation, lesser labour requirement	puddled transplanted rice (PTR)-Wheat	Bhullar et al. (2018)
India	Eggplant	Drip irrigation	Reduced water, electricity and fertilizer use, and increased returns	54% higher net returns	Narayanamoorthy et al. (2018)

Source: Reproduced from Malhi et al. (2021)

varieties and improved laser land levelling techniques contributed to increased household income, facilitating asset accumulation and investment in livestock breeds. The impressive rise in household income by 83% and 76% underscores the transformative impact of climate-smart technologies on rural livelihoods.

Furthermore, the table highlights other case studies, such as groundnut cultivation in the semi-arid tropics of India, where drought-tolerant varieties resulted in higher yields, reduced yield variability, and lower variable costs. The adoption of zero tillage in wheat production in Karnal, Haryana, enhanced production, reduced cultivation costs, and increased net income. Laser land levelling in north-western India for rice and wheat cultivation reduced irrigation time, increased yields, and decreased electricity charges.

These case studies collectively demonstrate the significance of climate-smart technologies in enhancing agricultural productivity and sustainability while improving economic outcomes for farmers. By leveraging innovative approaches tailored to specific crops and regions, farmers can optimize resource utilization, increase yields, reduce input costs, and ultimately improve their livelihoods.

Overall, the table highlights the potential of climate-smart technologies to transform agricultural systems, fostering resilience, resource efficiency, and economic prosperity for farming communities worldwide. The findings underscore the importance of promoting and supporting the widespread adoption of these technologies to address the challenges of food security, climate change, and sustainable development in the agricultural sector.

6.6   Climate Finance for Conserving Natural Capital in India

Climate change, largely regarded as the defining challenge of the twenty-first century, raises a complex web of environmental, economic, and societal issues. The anthropogenic act has resulted in a cascade of impacts, ranging from melting polar ice caps to intensifying extreme weather occurrences, food and water scarcity, and displacement of vulnerable communities (Weiskof et al., 2020). Recognizing the gravity of the situation, the international community has banded together under the auspices of the United Nations Framework Convention on Climate Change (UNFCCC) to forge agreements and mechanisms to mitigate the effects of climate change and assist societies in adapting to its unavoidable consequences.

Among the notable agreements that have resulted from this worldwide endeavour, the Paris Agreement of 2015 stands out as a symbol of united resolve. Its primary goal is to keep global average temperatures below 2°C over pre-industrial levels while pursuing initiatives to keep the increase below 1.5°C. To reach this aim, the agreement asks for a significant increase in climate funding for the transition to low-carbon, resilient economies.

The idea of establishing a dedicated fund to support climate action in developing countries gained traction in the lead-up to the 2009 United Nations Climate Change Conference in Copenhagen. The concept was crystallized in the Copenhagen Accord, which called for the mobilization of $100 billion annually in climate finance every year by 2020 (Seo, 2019). Subsequently, the Green Climate Fund was formally established during the 16th meeting of the Conference of the Parties to the UNFCCC in 2010 as a unique multilateral financial entity exclusively serving the UNFCCC. Its core mission is to direct investments, both public and private, towards low-emission and climate-resilient projects in assisting underprivileged and emerging economies with low-carbon development. It fills a key vacuum in climate finance by focusing on mitigation and adaptation initiatives (Brechin and Espinoza, 2017). The GCF's creation has been seen as a significant step forward in the global climate finance landscape. It was designed to serve as a centrepiece of the financial mechanism outlined in the Paris Agreement. The fund was envisioned as a way to operationalize the principle of "balance" in climate finance, ensuring that support for mitigation and adaptation was equitable and sufficient to meet the needs of developing countries(Carraro and Massetti, 2012).

Over the years, the GCF has evolved, refining its governance structure, funding mechanisms, and operational procedures. It has engaged a diverse array of stakeholders, from governments and multilateral development institutions to civil society organizations and the private sector, in a concerted effort to combat climate change. GCF utilizes a variety of financial mechanisms, such as grants, loans, and equity, emphasizing the need for international financial institutions to harmonize environmental and social criteria to fund projects aligned with climate goals. Bridging the financial gap is vital to ensure effective GHG emission reduction in developing countries (Ari and Isik, 2022). Notably, it disbursed a record €4.9 billion for climate-related projects in 2020 (Ahamer, 2021). Public climate finance for developing countries (mainly provided by bilateral and multilateral institutions and mainly targeted at mitigation) amounted to between US$35 and 49 billion per year in 2011 and 2012. A substantial part of these funds, however, resulted from relabelling existing activities rather than new commitments. The GCF has funded $539.28 million for environmental projects in India. These include projects like the $34.36 million NABARD project, which uses solar irrigation to improve water security and resilience in Odisha's tribal areas, and the $100 million credit line for solar rooftop installations in the commercial, industrial, and residential sectors. The $137 million Green Growth Equity Fund from FMO promotes large-scale mitigation initiatives, while the $43.42 million project from UNDP aims to improve climate resilience in coastal communities in India. The $24.5 million Avaana Sustainability Fund of SIDBI supports broad-based sustainability objectives. Furthermore, the $200 million investment made by MAAML in India's E-Mobility Financing Program highlights the role played by the private sector in mitigating climate change. Together, these initiatives demonstrate a thorough strategy involving both public and private, including small, medium, and large-scale activities in both the public and private sectors. These investments are significantly contributing to India's natural capital

Table 6.3 GCF Projects in India

Project name	Entity	Sector	Theme	Project size	FA financing
Ground Water Recharge and Solar Micro Irrigation to Ensure Food Security and Enhance Resilience in Vulnerable Tribal Areas of Odisha	NABARD	Public	Adaptation	Medium	34,357,000
Line of Credit for Solar rooftop segment for commercial, industrial and residential housing sectors	NABARD	Private	Mitigation	Medium	100,000,000
Enhancing climate resilience of India's coastal communities	UNDP	Public	Cross-cutting	Medium	43,418,606
Green Growth Equity Fund	FMO	Private	Mitigation	Large	137,000,000
India E-Mobility Financing Program	MAAML	Private	Mitigation	Large	200,000,000
Avaana Sustainability Fund	SIDBI	Private	Cross-cutting	Small	24,500,000
Total					**539,275,606**

by improving water management, encouraging renewable energy adoption, conserving coastal ecosystems, advancing green technologies, and developing sustainability ideas by enhancing resource efficiency, lowering carbon emissions, and increasing resilience to climate effects (Table 6.3).

According to the Global Environment Facility (GEF) approximately 142 projects, global and regional projects at a gross grant of around $1317.9 million are going on in India, and around 94 national projects are going on at a grant of $702,244,948. Most of these projects' focal areas are biodiversity ($91 million), climate change ($437 million), land degradation ($11 million) and chemical & waste ($26 million). From Table 6.4, we can see that most of the projects are allocated to climate change themes and biodiversity. Around 10 projects focused on different sectors at the same time with a grant of $114 million.

Table 6.4 GEF Projects in India

Focal area	No of project	GEF grant (In$)
Biodiversity	21	91,511,272
Climate change	54	437,110,263
Land degradation	3	11,721,411
Chemical waste	4	26,100,000
Mixed	10	114,390,927

6.7 National Policy and Programmes for CSA and Natural Capital Conservation

6.7.1 *National Innovation on Climate Resilient Agriculture (NICRA)*

The Indian Council of Agricultural Research (ICAR) is the organization in charge of the National Innovations in Climate Resilient Agriculture (NICRA), which was introduced in February 2011 with a budget of Rs. 350 crores. Its goal is to increase Indian agriculture's capacity to withstand climate change and variability while also producing more crops, animals, and fisheries. Advanced manufacturing methods and risk management technologies are developed and put into use to accomplish this.

The NICRA project has focused on developing climate-resilient technologies for diverse crops through state-of-the-art climate change research facilities across the country. It has also conducted district-level risk assessments for climate change's impact on Indian agriculture, resulting in the development of District Agriculture Contingency Plans by ICAR and NARS for 650 districts. Additionally, NICRA has established climate-resilient villages in 151 vulnerable districts, showcasing location-specific technologies.

6.7.2 *National Mission on Sustainable Agriculture (NMSA)*

Under the NAPCC, the NMSA implements various programmes such as Soil Health Card, Paramparagat Krishi Vikas Yojana, and Mission Organic Value Chain Development. The NMSA promotes sustainable practices, energy efficiency, natural resource conservation, and location-specific agronomic improvements.

6.7.3 *The National Adaptation Fund for Climate Change (NAFCC)*

It was established to provide assistance in adaptation to climate change in vulnerable Indian states and union territories. It funds projects in sectors like agriculture, with implementation across various states including Punjab, Himachal Pradesh, Odisha, and more.

6.7.4 *Climate Smart Village (CSV)*

CSV approach promotes the local-level implementation of CSA to enhance farmers' climate change adaptation capabilities. Multiple research organizations, including CGIAR centres, collaborate on CSV initiatives in India. Initially piloted in the Karnal district of Haryana and Vaishali district of Bihar, CSVs have expanded to include Punjab, Andhra Pradesh, and Karnataka districts.

6.7.5 *Pradhan Mantri Krishi Sinchayee Yojna (PMSKY)*

The programme focuses on water management and conservation in agriculture with the goal of increasing irrigation coverage and streamlining water consumption. "Har Khet Ko Paani" is the major catchphrase of this programme, which aims to increase water usage effectiveness. The "More crop per drop" programme offers complete water management, source generation, and channel distribution options.

The emphasis on "More crop per Drop" reflects India's comprehensive approach to water management in agriculture. PMSKY and MGNREGA have significantly contributed to water resource conservation, recharge, and utilization of groundwater. PMSKY includes the Micro Irrigation Fund, which prioritizes protective irrigation and interventions for water use efficiency (WUE) in a climate-conscious manner.

6.7.6 *Pradhan Mantri Fasal Bima Yojna (PMFBY)*

The PMFBY, implemented in India since the Kharif 2016 season, offers voluntary participation for both states/union territories and farmers. States/UTs can choose to participate based on their risk assessment and financial factors. As of 2020–2021, a total of 2,938.7 lakh farmer applications, covering an insured sum of Rs. 1,049,342 crores, have been enrolled in the scheme.

6.7.7 *Soil Health Card Scheme (SHC)*

SHC Scheme, initiated in February 2015, aims to facilitate farmers with comprehensive information about the nutrient status of their land based on soil tests. This enables farmers to make informed decisions regarding fertilizer use for enhanced productivity. The Government of India set a target to issue 10.48 crores of SHCs since the inception of the scheme.

6.7.8 *National Water Mission (NWM)*

An ambitious step was undertaken to enhance WUE by 20%, encompassing the agricultural sector, and to promote Integrated Water Resource Management (IWRM) practices. The aim is to safeguard water sources and curtail waste. This initiative strives to optimize water utilization, particularly in the agricultural domain, to ensure sustainable water management and mitigate the challenges of water scarcity.

6.7.9 *Paramparagat Krishi Vikas Yojna (PKVY)*

It is an extension of Soil Health Management scheme, introduced in 2015 under the NMSA. Its goal is to support and promote organic farming by adopting a village-by-cluster strategy that improves soil health.

6.7.10 *Biotech-KISAN*

The Biotech-KISAN programme, initiated in 2017, fosters collaboration between scientists and farmers to drive agricultural innovation. With 146 hubs established across India's aspirational districts and agro-climatic zones, the programme has benefited over two lakh farmers, boosting their income and agrarian output while generating over 200 rural entrepreneurs. Notably, the programme's implementation has resulted in significant positive impacts, including increased food grain production, prevention of forest area conversion, reduced GHG emissions, and remarkable growth in the country's milk output through the implementation of livestock-related laws prioritizing agribusiness ecosystem marketing and animal health.

6.8   Policy and Institutional Barriers to CSA and Natural Capital Conservation

CSA and natural capital conservation are vital to bolster agricultural resilience, reduce GHG emissions, and advance sustainable development. Nevertheless, the successful execution of CSA and natural capital conservation practices in India encounter numerous policy and institutional obstacles.

- Lack of coordination between different government agencies. Several government agencies are responsible for agriculture, climate change, and natural resources. However, there is often a lack of coordination between these agencies, which can lead to duplication of efforts and a lack of progress on CSA and natural capital conservation.
- Weak enforcement of laws and regulations. Some laws and regulations are designed to promote CSA and natural capital conservation. However, these laws and regulations are often not enforced effectively, which can lead to a decline in natural resources and a loss of agricultural productivity.
- Need more adequate access to finance. Farmers often lack the financial resources to adopt CSA practices and conserve natural resources. This is due to several factors, including low agricultural incomes, high interest rates, and a lack of access to credit.
- Limited awareness of CSA and natural capital conservation. Many farmers are not aware of the benefits of CSA and natural capital conservation. This is due to a lack of education and outreach programmes.

These barriers can be overcome through several policy and institutional reforms. These reforms include:

- Strengthening coordination between government agencies. This can be done by creating a single coordinating body for CSA and natural capital conservation.
- Strengthening enforcement of laws and regulations. This can be done by increasing the number of inspectors and providing them with the necessary resources to enforce the law.
- Providing farmers with access to finance. This can be done by providing farmers with low-interest loans and creating microfinance institutions specializing in lending to farmers.
- Raising awareness of CSA and natural capital conservation. This can be done through education and outreach programmes.

These reforms are essential for successfully implementing CSA and natural capital conservation in India. By addressing these barriers, India can contribute to the global effort to mitigate climate change and adapt to its effects.

6.9   Conclusion

Human activities have significantly altered the natural ecosystem, posing a global threat that necessitates the adoption of sustainable development principles. Sustainable development involves a holistic approach that integrates social, environmental, and economic considerations to address these challenges while safeguarding the future. This requires robust policy formulation and increased awareness.

Concerns over water scarcity, declining agricultural productivity, and depleting resources underscore the need for sustainable farming practices. Diversifying cropping systems through conservation agriculture-based intensification offers a promising approach to enhance farm output and address declining water productivity simultaneously. CA practices, such as zero tillage and minimizing SOC losses, enhance climate change resilience and mitigate soil degradation. Scientists should evaluate resource-conserving technologies grounded in sound scientific principles to reduce atmospheric emissions of CO_2, CH_4, and N_2O.

While modern agriculture has achieved exceptional yields, its heavy reliance on chemical inputs and conventional energy sources has raised significant concerns. The considerable costs and adverse impacts on public and environmental health and soil degradation have become increasingly apparent. Given soil's critical role in the agricultural production system, soil health has to be reevaluated to ensure sustainable agricultural output. To lessen the possible effects of climate change on agriculture, the evaluation and distribution of climate-resilient soil management techniques are crucial. A significant source of the greenhouse gases CH_4, N_2O, and CO_2 that contribute to global warming because of continuing climate change is agriculture.

In summary, transitioning towards sustainable development is crucial to address the global threat posed by human-induced changes to the natural ecosystem. Implementing robust policies and fostering awareness will be essential. Embracing conservation measures in agriculture, adopting resource-conserving technologies, and prioritizing soil health in the face of climate change is imperative to ensure long-term agricultural sustainability and reduce GHG emissions.

References

Aggarwal, P. K., Alduchov, O. A., Froehlich, K. O., Araguas-Araguas, L. J., Sturchio, N. C. and Kurita, N. (2012) 'Stable isotopes in global precipitation: A unified interpretation based on atmospheric moisture residence time.' *Geophysical Research Letters*, 39(11).

Ahamer, G. (2021) 'International financial Institutions ask to contribute to climate protection', *Finance: Theory and Practice*, 25(4), pp. 6–23. https://doi.org/10.26794/2587-5671-2020-25-4-6-23

Antle, J. M. and Capalbo, S. M. (2010) 'Adaptation of agricultural And food systems to climate change: An economic And policy perspective', *Applied Economic Perspectives and Policy*, 32, pp. 386–416.

Ari, I. and Isik, M. (2022) Assessing the performance of the developing countries for the utilization of the green climate fund. *Frontiers in Climate*, 4. https://doi.org/10.3389/fclim.2022.813406

Arora, M., Goel, N. K. and Singh, P. (2005) 'Evaluation of temperature trends over India/Evaluation de tendances de temperature en Inde', *Hydrological Sciences Journal*, 50, pp. 81–93.

Aryal, J. P., Mehrotra, M. B., Jat, M. L. and Sidhu, H. S. (2015) 'Impacts of laser land leveling in rice–wheat systems of the north–western indo-gangetic plains of India,' *Food Security*, 7(3), pp. 725–738.

Aryal, J. P., Sapkota, T. B., Jat, M. L. and Bishnoi, J. K. (2015b) 'On-farm economic and environmental impact of zero-tillage wheat: A case of north-west India', *Experimental Agriculture*, 51, pp. 1–16.

Asner, G. P., Elmore, A. J., Olander, I. P., Martin, R. E. and Harris, A. T. (2004) 'Grazing systems, ecosystems responses, and global change', *Annual Review of Environmental Resources*, 29, pp. 261–299.

Bentley, D., Hegarty, R. S. and Alford, A. R. (2008) 'Managing livestock enterprises in Australia's extensive rangelands for greenhouse gas and environment outcomes: A pastoral company perspective', *Australian Journal of Experimental Agriculture*, 48(1–2), pp. 60–64. https://doi.org/10.1071/EA07210

Berndes, G., Hoogwijk, M. and Van Den Broek, R. (2003) 'The contribution of biomass in the future global energy supply: A review of 17 studies', *Biomass and Bioenergy*, 25(1), pp. 1–28. https://doi.org/10.1016/S0961-9534(02)00185-X

Bhatt, R., Arora, S. and Chew, C. C. (2016) 'Improving irrigation water productivity using Tensiometers,' *Journal of Soil and Water Conservation*, 15(2), pp. 120–124.

Bhatt, R. and Kukal, S. S. (2014) 'Resource conservation techniques for mitigating the water scarcity problem in the Punjab, India – A review', *International Journal of Earth Sciences and Engineering*, 7(1), pp. 309–316.

Bhullar, M. S., Singh, S., Kumar, S. and Gill, G. (2018) 'Agronomic and economic impacts of direct seeded rice in Punjab', *Agricultural Research Journal*, 55, pp. 236–242.

Birthal, P. S., Nigam, A. N., Narayanan, A. V. and Kareem, K. A. (2012) 'Potential economic benefits from the adoption of improved drought-tolerant groundnut in India', *Agricultural Economics Research Review*, 25, pp. 1–14.

Bolin, B., Crutzen, P. J., Vitousek, P. M., Woodmansee, R. G., Goldberg, E. D. and Cook, R. B. (1983) 'The major biogeochemical cycles and their interactions,' *Scientific Committee on Problems of the Environment (SCOPE)*, of the International Council of Scientific Unions (ICSU) by Wiley, University of California.

Brechin, S. R. and Espinoza, M. I. (2017) 'A case for further refinement of the green climate fund's 50:50 ratio climate change mitigation and adaptation allocation framework: Toward a more targeted approach', *Climatic Change*, 142(3–4), pp. 311–320. https://doi.org/10.1007/s10584-017-1938-8

Carraro, C. and Massetti, E. (2012) 'Beyond Copenhagen: A realistic climate policy in a fragmented world', *Climatic Change*, 110(3–4), pp. 523–542. https://doi.org/10.1007/s10584-011-0125-6

Chambers, R. (1991) Scientist or resource-poor farmer: Whose knowledge counts? *Crop Protection for Resource Poor Farmers*, 1–15. https://cris.brighton.ac.uk/ws/

portalfiles/portal/4756051/WHOSE%20KNOWLEDGE%20COUNTS_CJ-Davies%20PhD%20FINAL%20JUNE16.pdf

Dadhich, R. K., Meena, R. S., Reager, M. L. and Kansotia, B. C. (2015) 'Response of bio-regulators to yield And quality of Indian mustard (*Brassica juncea* L. Czernj. and Cosson) under different irrigation environments', *Journal of Applied and Natural Science*, 7(1), pp. 52–57. https://doi.org/10.31018/jans.v7i1.562

Dadhwal, V. K., Parihar, J. S. and Sudhakar, S. (2014) Role of forests in climate change mitigation. In *Forest ecosystems* (pp. 389–403). Paris: Springer.

Falloon, P. and Betts, R. (2010) 'Climate impacts on European agriculture and water management in the context of adaptation and mitigation-the importance of an integrated approach', *Science of the Total Environment*, 408(23), pp. 5667–5687. https://doi.org/10.1016/j.scitotenv.2009.05.002

Falloon, P., Falloon, P., Powlson, D. and Smith, P. (2004) 'Managing field margins for biodiversity and carbon sequestration: A great Britain case study', *Soil Use and Management*, 20(2), pp. 240–247. https://doi.org/10.1111/j.1475-2743.2004.tb00364.x

Food and Agriculture Organization of the United Nations (FAO) reports on "Natural Capital Impacts in Agriculture" (2015) https://www.fao.org/nr/sustainability/natural-capital.

Fraser, E. D. G., Simelton, E., Termansen, M., Gosling, S. N. and South, A. (2013) 'Vulnerability hotspots': Integrating socio-economic and hydrological models to identify where cereal production May decline due to climate change-induced drought', *Agricultural and Forest Meteorology*, 170, pp. 195–205.

Gomez, B. (2005) *Degradation of vegetation and agricultural productivity due to Natural disasters and land use strategies to mitigate their impacts on agriculture, rangelands and forestry: Natural disasters and extreme events in agriculture-impacts and mitigation*. Berlin: Springer Verlag, p. 259–276.

Humphreys, E., Kukal, S. S., Christen, E. W., Hira, G. S. and Sharma, R. K. (2010) 'Halting the groundwater decline in north-west India—which crop technologies will be winners?' *Advances in Agronomy*, 109, pp. 155–217.

Huntington, T. (2006) Available water capacity and soil organic matter. *Encyclopedia of soil science*. 2nd edition. Boca Raton: Taylor & Francis.

International Food Policy Research Institute. (2017) *2017 Global Food Policy Report*. Washington, DC: International Food Policy Research Institute. https://doi.org.10.2499/9780896292529

IPCC (Intergovernmental Panel on Climate Change). (2007) *Climate change 2007: Impacts, adaptation and vulnerability. Contribution of working group II to the fourth assessment report of the intergovernmental panel on climate change*. Cambridge, UK: Cambridge University Press; p. 976.

Jalota, S. K. and Arora, V. K. (2002) 'Model-based assessment of water balance components under different cropping systems in north-west India', *Agricultural Water Management*, 57, pp. 75–87.

Jalota, S. K., Khera, R., Arora, V. K. and Beri, V. (2007) 'Benefits of straw mulching in crop production: A review', *Journal of Research Punjab Agricultural University*, 44, pp. 104–107.

Jena, P. R. (2019) 'Can minimum tillage enhance productivity? Evidence from smallholder farmers in Kenya', *Journal of Cleaner Production*, 218, pp. 465–475. https://doi.org/10.1016/j.jclepro.2019.01.278.

Kaye, N. M., Mason, S. C., Galusha, T. D. and Mamo, M. (2007) 'Nodulating and non-nodulating soybean rotation influence on soil nitrate–nitrogen and water, and sorghum yield', *Agronomy Journal*, 99, pp. 599–606.

Kumari, S. (2017) 'Effects of nitrogen levels on anatomy, growth, and chlorophyll content in sunflower (*Helianthus annuus* L.) Leaves', *Journal of Agricultural Science*, 9(8), pp. 208–219.

Lal, R. (2005) 'Forest soils and carbon sequestration', *Forest Ecology and Management*, 220(1–3), pp. 220, 242–258.

Latif, A., Shakir, A. S. and Rashid, M. U. (2013) 'Appraisal of economic impact of zero tillage, laser land levelling and bed-furrow intervention in Punjab, Pakistan', *Pakistan Journal of Engineering and Applied Sciences*, 13, p. 6581.

Malhi, G. S., Kaur, M. and Kaushik, P. (2021) 'Impact of climate change on agriculture and its mitigation strategies: A review', *Sustainability*, 13(3), p. 1318. MDPI AG. Retrieved from http://dx.doi.org/10.3390/su13031318

Meena, H., & Meena, R. S. (2017). Assessment of sowing environments and bio-regulators as adaptation choice for clusterbean productivity in response to current climatic scenario. Bangladesh J Bot, 46(1), 241–244.

Mishra, A. K., Khanal, A. R. and Pede, V. Economic and resource conservation perspectives of direct seeded rice planting methods: Evidence from India. In Proceedings of the Agricultural and Applied Economics Association's 2017 AAEA Annual Meeting, Chicago, IL, USA, 30 July–2 August 2016.

Motha, R. P. (2007) 'Development of an agricultural weather policy', *Agricultural and Forest Meteorology*, 142(2–4), pp. 303–313. https://doi.org/10.1016/j.agrformet.2006.03.031

Narayanamoorthy, A., Bhattarai, M. and Joshi, P. (2018) 'An assessment of the economic impact of drip irrigation in vegetable production in India', *Agricultural Economics Research Review*, 31, pp. 105–112.

Narayanamoorthy, A. and Devika, N. (2017) 'Economic and resource impacts of drip method of irrigation on okra cultivation: Analysis of field survey data', *Journal of Land and Rural Studies*, 6, pp. 15–33.

NASA Earth Observatory. Goddard Space Flight Centre United States. Available online: www.earthobservatory.nasa.gov (accessed on 15 May 2020).

Ogada, M. J., Rao, E. J. O., Radeny, M., Recha, J. W. and Solomon, D. (2020) Cllimate-smart agriculture, household income and asset accumulation among smallholder farmers in Nyando basin of Kenya. *World Development Perspectives*, 18, p. 100203. https://doi.org/10.1016/j.wdp.2020.100203

Pampolino, M. F., Manguiat, I. J., Ramanathan, S., Gines, H. C., Tan, P. S., Chi, T. T. N., Rajendran, R. and Buresh, R. J. (2007) 'Environmental impact and economic benefits of site-specific nutrient management (SSNM) in irrigated rice systems', *Agricultural Systems*, 93, pp. 1–24.

Parmesan, C. (2006) 'Ecological and evolutionary responses to recent climate change,' *Annual Review of Ecology, Evolution, and Systematics*, 37, 637–669. https://doi.org/10.1146/annurev.ecolsys.37.091305.110100

Post, D. M., Pace, M. L. and Hairston Jr, N. G. (2000) 'Ecosystem size determines food-chain length in lakes,' *Nature*, 405(6790), pp. 1047–1049.

Purakayastha, T. J., Kumari, S. and Pathak, H. (2015) 'Characterisation, stability, and microbial effects of four biochars produced from crop residues', *Geoderma*, 239, pp. 293–303.

Rafferty, N. E. and Ives, A. R. (2010) Effects of experimental shifts in flowering phenology on plant-pollinator interactions. *Ecology Letters*, 14(1), pp. 14, 69 74. https://doi.org/10.1111/j.1461-0248.2010.01557.x

Rai, P. K., Rai, N. K. and Kumar, P. (2015) Role of pollinators in crop production and food security. In *Bees in sustainable agriculture*. Springer, pp. 161–175, https://www.researchgate.net/publication/294287339_Role_of_Pollinators_in_Sustainable_Farming_and_Livelihood_Security

Ram, K. and Meena, R. S. (2014) 'Evaluation of pearl millet and mungbean intercropping systems in arid region of Rajasthan (India)', *Bangladesh Journal of Botany*, 43(3), pp. 367–370.

Ric Powlson, D. S., Christian, D. G., Falloon, P. and Smith, P. (2001) Biofuel crops: Their potential contribution to decreased fossil carbon emissions and additional environmental benefits. *Aspects Applied Biology*, pp. 65, 289–294.

Rosenzweig, C. (2007) Executive summary. In: Parry ML et al (eds.) *Chapter 1: Assessment of observed changes and responses in natural and managed systems. climate change 2007: Impacts, adaptation and vulnerability: Contribution of working group II to the fourth assessment report of the intergovernmental panel on climate change.* Cambridge: Cambridge University Press (CUP)

Seo, S. N. (2019) Economic questions on global warming during the trump years. *Journal of Public Affairs*, 19(1). https://doi.org/10.1002/pa.1914

Shiferaw, B., MenaleKassie, M., Jaleta, M. and Yirga, C. (2014) 'Adoption of improved wheat varieties and impacts on household food security in Ethiopia', *Food Policy*, 44, pp. 272–284.

Singh, V. K., Shukla, A. K., Singh, M. P., Majumdar, K., Mishra, R. P., Rani, M. and Singh, S. K. (2015) 'Effect of site-specific nutrient management on yield, profit and apparent nutrient balance under pre-dominant cropping systems of upper gangetic plains', *Indian Journal of Agricultural Sciences*, 85, pp. 335–343.

Smith, K., Cumby, T., Lapworth, J., Misselbrook, T. and Williams, A. (2008) 'Natural crusting of slurry storage as an abatement measure for ammonia emissions on dairy farms', *Biosystems Engineering*, 97, pp. 464–471.

Stroosnijder, L. (2007) 'Rainfall and land degradation', in Sivakumar, M. V. K. and Ndegwa, N. (eds.) *Climate and land degradation.* Germany: Springer-Heidelberg, pp. 167–95.

Thornton, P.K.; Herrero, M. 2010. The inter-linkages between rapid growth in livestock production, climate change, and the impacts on water resources, land use, and deforestation. Background paper for the 2010 World Development Report. World Bank Policy Research Working Paper 5178. Washington DC: World Bank.https://cgspace.cgiar.org/items/a4bd9553-67b5-4431-8164-7665466413ab

Tol, R. S. J. (2013) 'The economic impact of climate change in the 20th and 21st centuries', *Climatic Change*, 117, pp. 795–808.

Tripathi, R. S., Raju, R. and Thimmappa, K. (2013) 'Impacts of zero tillage on economics of wheat production in Haryana', *Agricultural Economics Research Review*, 26, pp. 101–108.

UNEP (2008) Annual report. https://wedocs.unep.org/bitstream/handle/20.500.11822/7742/-UNEP%202008%20Annual%20Report-2009837.pdf?sequence=3&%3BisAllowed=

United Nations Office for Disaster Risk Reduction: 2018 *Annual report.* https://www.undrr.org/publication/united-nations-office-disaster-risk-reduction-2018-annual-report.

Varvel, G. E. and Wilhelm, W. W. (2003) 'Soybean nitrogen contribution to corn and sorghum in western corn belt rotations', *Agronomy Journal*, 95, pp. 1220–1225.

Verma, S. K., Singh, S. B., Prasad, S. K., Meena, R. N. and Meena, R. S. (2015) 'Influence of irrigation regimes and weed management practices on water use and nutrient uptake in wheat (Triticum aestivum L. Emend. Fiori and Paol.),' *Bangladesh Journal of Botany*, 44(3), pp. 437–442.

Wagan, S. A., Memon, Q. U. A., Wagan, T. A., Memon, I. H. and Wagan, A. (2015) 'Economic analysis of laser land levelling technology water use efficiency and crop productivity of wheat crop in Sindh, Pakistan', *Journal Environmental Earth Sciences*, 5, pp. 21–25.

Wallace, J. S. and Gregory, P. J. (2002) 'Water resources and their use in food production systems,' *Aquatic Sciences*, 64, pp. 363–375.

Weiskopf, Sarah R., Madeleine A. Rubenstein, Lisa G. Crozier, Sarah Gaichas, Roger Griffis, Jessica E. Halofsky, Kimberly JW Hyde et al. "Climate change effects on biodiversity, ecosystems, ecosystem services, and natural resource management in the United States." Science of the Total Environment 733 (2020): 137782.

7 Impact of CSA Practices on Household Income and Agricultural Yield

7.1 Introduction

From 2016 to 2021, natural disasters such as cyclones, flash floods, floods and landslides resulted in crop damage on more than 36 million hectares of land, causing financial losses of around \$3.75 billion for farmers in India. It is expected that the annual damages caused by river flooding in the country will likely increase by about 49% if the temperature increases by 1.5°C, while the damage caused by cyclones is projected to increase by 5.7% (The Hindu, 2022). So, at this highly vulnerable stage, climate-smart agricultural (CSA) practices have the potential to increase productivity, strengthen resilience to climatic shocks and reduce greenhouse gas emissions (FAO 2011). CSA emphasizes food and income security in the era of advancing climate change and variability through the development of resilient agricultural production systems (Vermeulen et al., 2012; Lipper et al., 2014).

CSA practices benefit both private and public entities by increasing the farmers' productivity and income. They also maintain food security and eradicate poverty among rural farmers. As a public benefit, CSA tends to mitigate climate change in the environment by reducing the release of greenhouse gas emissions (Pretty, 2008; Branca et al., 2011). Many studies have established the triple-win nature of CSA adoption, that is production, mitigation and income in developing and developed countries (Xiong et al., 2014; Challinor et al., 2014; Mungai et al., 2016; Makate et al., 2017; Lan et al., 2018). CSA practises have been sustainable and environment-friendly and have enhanced both yield and income (Wekesa et al., 2018; Makate et al., 2018; Ali et al., 2018).

In India, there have been a limited number of research studies that have examined the effects of CSA on the well-being of households. These studies have found that implementing practices such as the use of improved crop varieties, laser land levelling, and zero tillage can lead to an overall increase in production in the rice-wheat system in the Indo-Gangetic plain region of the country. The adoption of these CSA practices has a significant impact on the reduction of the cost of production (Khatri-Chhetri et al., 2016). Conservation agriculture (CA) and improved livestock husbandry have improved food security for large and medium-scale farmers in Bihar, India (Lopez-Ridaura et al., 2018).

DOI: 10.4324/9781003290445-7

Measures to conserve soil and water have had a positive effect on both farm productivity and income in the semi-arid region of Bundelkhand in central India (Choudhary et al., 2022). Soil smart practices such as regular soil bund reduce the chances of downside risk, i.e., crop failure (Kumar et al., 2020). Adopting crop-smart and water-smart strategies meets several objectives of the SDGs (Sustainable Development Goals) (Qureshi et al., 2022). Due to climate change, catastrophic weather events such as floods and droughts are becoming commonplace, greatly increasing the unpredictability of agricultural output in Odisha (Mishra et al., 2016). Due to regular shocks in agriculture, crop loss, poor harvests and unpaid bank credit, a few farmers have even committed suicide in Odisha (Mohanty and Lenka, 2019) One of the major issues with rice production, particularly in the rain-fed lowland areas of Odisha, is flash floods that wash away rice plants for 10–15 days. The paddy crop fails due to irregular rainfall and delayed southwest monsoon in the inland districts of Odisha. However, the farmers of Odisha are changing the nature of agricultural production. To cope with climate change, they are switching from conventional farming to CSA practices (Tanti et al., 2022). Farmers of Odisha are gradually adopting a basket of CSA practices such as rescheduling planting, crop rotation, crop diversification, drought-resistant seeds and smart soil practices (Sahu and Mishra, 2013). No significant micro studies have captured the impact of the adoption of CSA practices in the vulnerable regions of Odisha. It has been proposed in many empirical studies that researchers should investigate the difference in income and yield that exists between small-scale farmers who adopt CSA practices and those who do not adopt them. This could provide information about the benefits derived from taking steps to adapt to climate change. The main research question addressed by this chapter is: What is the impact of CSA adoptions on paddy yield and farm income in rural Odisha?

Given the above background, this research primarily focuses on assessing the effects of implementing CSA practices on the yield of paddy and the income of households, which provides a valuable addition to the existing literature.

7.2   Materials and Methods

7.2.1   Study Area

Odisha, one of the states in eastern India, was selected as the area of study. Odisha's economy is based mostly on agriculture and farming. About 83% of its population is from a rural background, and about 61.8% of its 17.5 million people work in the agricultural sector, with 18% of the state's GDP coming from the same. The samples from three districts have been drawn for the impact analysis. Four hundred ninety-four household observation data from the three study districts of Odisha is used to assess the impact of CSA on income and yield. The description of sampling and data collection is discussed in Chapter 4. The following sections discuss data description and the econometrics method used for the impact analysis.

7.2.2　*Data Description*

7.2.2.1　*Endogenous Variable/Adoption Variable*

We required a specific data set to do the econometrics analysis: we took two endogenous variables, which were different adoption strategies followed by the farmers. Specifically, two CSA practices such as crop rotation and integrated soil management practices are chosen for this analysis. We have selected these two CSA adoption practices based on the extension officers' recommendations and preliminary observations from the field. The endogenous variables had two groups of households: one was the adopter group and the other was the non-adopter group. The endogenous variables are binary: 1 holds if the household is an adopter of the specific CSA practice; otherwise, 0.

7.2.2.2　*Income and Yield*

Total farm income of the household in a year was taken as a dependent variable in the study. We have calculated total farm income by adding the total net income self-reported by the household from the Kharif and the Rabi seasons. We have asked the household to report the total net income (total agriculture income in a year – total expenses incurred for agricultural operations). We have not included the secondary or off-farm income they receive from various other sources. The significant income received from paddy, maize, cotton and vegetables per annum was included. To deal with the outliers in the sample, to make the data distribution smooth and to interpret the coefficient in percentage terms, the dependent variable has been transformed by using its natural logarithm. Our second dependent variable is the total paddy yield. We have arrived at the total paddy yield by dividing the total paddy production in a year by the total cultivated land size. The cultivated land size is the summation of the total owned land holding and the land taken on lease for cultivation.

7.3　Results and Discussion

7.3.1　*Descriptive Statistics*

Table 7.1 shows the descriptive statistics of dependent, endogenous, instrumental and controlled variables. The total annual agriculture income from various crops reported by households is presented in rupees. The average farm income earned by the farmers in the study region is INR 92,810.192. The highest income reported is INR 473,000, and the lowest is INR 2,400. The uneven land distribution is an important causal factor for this income gap. The average paddy productivity is reported at 16.5 quintals per acre. The maximum paddy productivity reported is 31.2 quintals per acre, while the minimum is 7.5 quintals per acre. Farmers who use improved varieties of seeds and have suitable land for cultivation are likely to get higher productivity than those who use traditional varieties of seeds and depend on rainfed agriculture. Around 62% of the farmers undertake integrated soil management practices, while 59% have adopted crop rotation.

Table 7.1 Descriptive Statistics

Variables		Mean	Std. Dev.	Min	Max
Dependent	Net agriculture income (log)	11.021	0.964	7.783	13.067
	Net agri-income/year	92,810.19	85,673.87	2,400	473,000
	Paddy productivity	16.513	6.464	7.5	31.25
Endogenous	Crop diversification	0.311	0.462	0	1
	Crop rotation	0.587	0.493	0	1
	Integrated soil management	0.617	0.487	0	1
	Improved variety seeds	0.784	0.411	0	1
Instrumental	Distance to extension office	13.334	7.418	2.5	30
	Multiple adopter group 1	64.712	33.694	0	100
Controlled	Total area	2.872	2.223	0	8.75
Variables	Farming experience	25.603	12.79	1	65
	Age	50.719	11.687	18	82
	Govt extn	0.703	0.458	0	1
	Education	7.788	5.37	0	17
	Credit from cooperative	0.462	0.499	0	1
	Access to subsidies	0.281	0.45	0	1
	Access to electricity	0.226	0.418	0	1
	Access to diesel	0.138	0.345	0	1
	Access to kerosine	0.038	0.019	0	1
	Multiple energy	0.165	0.375	0	1
	Mechanization index	−4.01e	1.3779	−1.163	5.628
	Kendrapara	0.293	0.456	0	1
	Balangir	0.21	0.408	0	1
	Mayurbhanj	0.497	0.501	0	1

Note: Total sample size: 494.

Seasonal or annual crop rotation entails the switching of crops in the field. It is an essential component of CSA since it helps maintain soil health, control pests and weeds, and maintain soil organic matter. The farmers from the Balangir district follow rice–vegetable, rice–oilseeds, maize–pulse/oilseeds and fibre–pulses systems of crop rotation annually. Farmers follow the crop sequence, such as jute–rice–pulses and rice–green gram/black gram/groundnut in the Kendrapada district. The rice–mustard/linseed/Bengal gram/safflower/black gram/lentil/green gram system of crop rotation is followed annually. Around 31% of the farmers in the study district followed crop diversification. Paddy is a dominant crop in the study districts, but due to climate uncertainty and market conditions, paddy production and profit are not up to the mark. The farmers have adopted soil management measures such as gypsum application, enhancing the height of field bunds, mulching and crop rotation. Shrubs are planted, and stone bunds are constructed along the fence of the farm plots to restrict soil erosion.

The average distance from the sampled village to the Block Agriculture Extension Office reported is 13 km. The shortest distance covered is 2.5 km. The longest distance to reach the extension office is 30 km. It is a two-way process when farmers visit the extension office to get information about different schemes and programmes. Farmers also visit the office to get subsidies for seeds and other benefits given by the government. The block agriculture staff also visits the farmers. We learnt that the village-level agricultural worker visits the village and the fields twice or thrice per week. The second instrumental variable used in the model is the percentage of multiple adopters in a village. An average of 64% of farmers follow multiple adoption practices. They practise at least two adoptions. This instrument will act as a peer effect in the model and encourage a maximum number of farmers to undertake multiple adoptions, thus triggering adoption by other farmers.

Farmers reported that 70% of them get access to extension services within the sample. Male-headed households constituted our sample, with their mean age being 50.7 years, and, on average, there were five members in a household. Almost seven years of education had been attained by the family heads. Also, the mean cultivable land size held within the sample was 2.8 acres, where an average of 19% of the land was irrigated, and 81% of the land depended on rain-fed agriculture. About 59% of the land had fertile soil. These lands have black clay and loamy soil, which can hold water and moisture. The black soil in the Balangir district is suitable for cotton cultivation.

7.3.2   *Result of PSM*

7.3.2.1   *Impact of Crop Rotation on Yield and Income*

ATT for total agriculture income relative to the non-adopter group is statistically positively significant. The difference in mean of total agricultural income is 0.424 (Radius), 0.455 (Kernel), 0.427 (5-nearest neighbour) according to the matching method. Likewise, adopters and non-adopters of crop rotation practices have a positive yield difference per acre. On average, there is a 1.98–2.54 quintal difference in the yield per acre if a farmer has adopted crop rotation practices. The result is significant for the all-matching method. The balancing test in Table 7.2 shows a proper balance of covariates after matching.

Table 7.2 Average Treatment Effect for Yield and Total Agriculture Income

Outcome variables	Matching methods	Crop rotation	
		ATT	*SE*
Yield per acre	Radius	1.98***	0.452
	Kernel	2.02***	0.459
	5-Nearest neighbour	2.54***	0.6022
Total agricultural income	Radius	0.424***	0.079
	Kernel	0.455***	0.822
	5-Nearest neighbour	0.427***	0.116

*Significant at 10% level; **Significant at 5% level; ***Significant at 1% level

Table 7.3 Impact of Covariates Balancing Before and After Matching

Adoption practice	Indicators of covariate balancing	Before matching	After matching
Crop rotation	Pseudo R²	0.093	0.006
	LR chi²	61.87	4.62
	a p-value of log-likelihood	0.000	0.990
	Median absolute bias	12.7	3.6
	Total % bias reduction (%)	75.1	18.1

The overlapping graph also indicates an appropriate balance between adopters and non-adopters (Table 7.3 and Figure 7.1).

7.3.2.2   *Impact of Crop Diversification on Yield and Income*

Crop diversification helps farmers to diversify the risk of crop loss. Framers grow multiple crops to compensate for recovering the cropless. Crop diversification doesn't have a significant impact on the net agricultural income of the farmers. Crop diversification has a significantly positive impact on the yield. The adoption of crop diversification has 1.6–2.24 quintals of yield per acre. The balancing table and graph show that the data is balanced after matching. The covariates of adopters and non-adopters have been balanced accurately (Tables 7.4 and 7.5 and Figure 7.2).

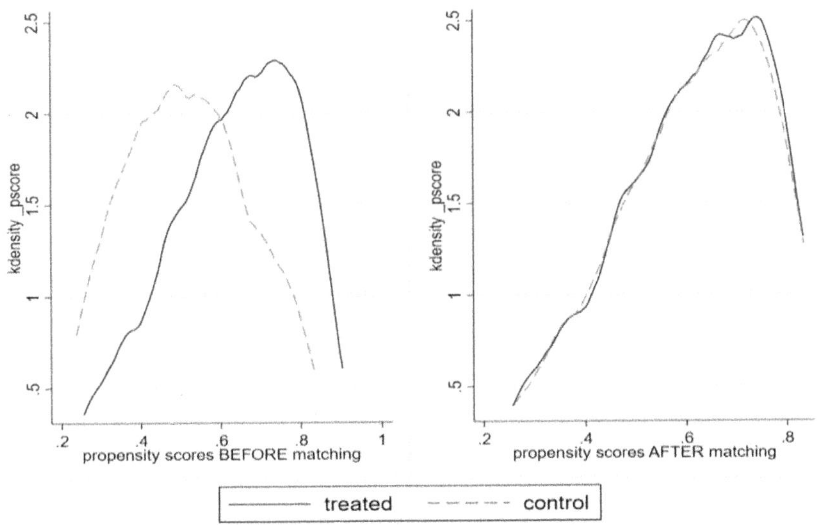

Figure 7.1 Propensity Score Overlapping between Treat and Control Groups, Before and After Matching for Crop Rotation.

Table 7.4 Average Treatment Effect for Yield and Total Agriculture Income

Outcome variables	Matching methods	Crop diversification	
		ATT	SE
Yield per acre	Radius	1.75***	0.452
	Kernel	1.69***	0.484
	5-Nearest neighbour	2.24***	0.664
Total agricultural income	Common radius calliper	0.033	0.104
	Kernel	0.025	0.109
	5-Nearest neighbour	0.180	0.137

Significant at 10% level; **Significant at 5% level; ***Significant at 1% level

Table 7.5 Impact of Covariates Balancing Before and After Matching

Adoption practice	Indicators of covariate balancing	Before matching	After matching
Crop rotation	Pseudo R²	0.049	0.004
	LR chi²	29.85	1.84
	p-value of log-likelihood	0.008	1.000
	Median absolute bias	7.0	3.1
	Total % bias reduction (%)	53.4*	15.4

Significant at 10% level; **Significant at 5% level; ***Significant at 1% level

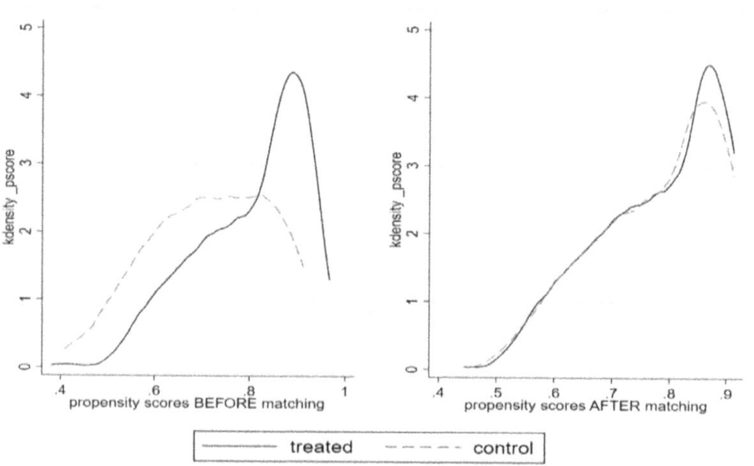

Figure 7.2 Propensity Score Overlapping between Treat and Control Groups, Before and After Matching for Soil Treatment.

7.3.2.3   Impact of Improved Variety Seeds on Yield and Income

A large number of farmers have shifted from the traditional variety of seeds to the improved variety of seeds. The improved variety of seeds helps to cope with the adverse impact of climate change. The farmers of this region were using drought and flood-resistance seeds. Farmers use early-maturity variety seeds due to the lack of irrigation facilities and uneven rainfall in the region. The farmers from the coastal region use flood-resistant seeds. Due to the lack of ditches and proper drains, farmers often face flash floods. The farmers face big losses due to crop damage during the harvesting period. Quintals between adopters and non-adopters of an improved variety of seeds (Tables 7.6 and 7.7 and Figure 7.3).

Table 7.6 Average Treatment Effect for Yield and Total Agriculture Income

Adoption practice	Indicators of covariate balancing	Before matching	After matching
Improve variety seeds	Pseudo R^2	0.082	0.017
	LR chi^2	41.87	15.58
	p-value of log-likelihood	0.000	0.340
	Median absolute bias	18.2	6.6
	Total % bias reduction (%)	73.8*	31.0*

Significant at 10% level; **Significant at 5% level; ***Significant at 1% level

Table 7.7 Impact of Covariates Balancing Before and after Matching

Adoption practice	Indicators of covariate balancing	Before matching	After matching
Improve variety seeds	Pseudo R^2	0.082	0.017
	LR chi^2	41.87	15.58
	p-value of log-likelihood	0.000	0.340
	Median absolute bias	18.2	6.6
	Total % bias reduction (%)	73.8*	31.0*

Significant at 10% level; **Significant at 5% level; ***Significant at 1% level

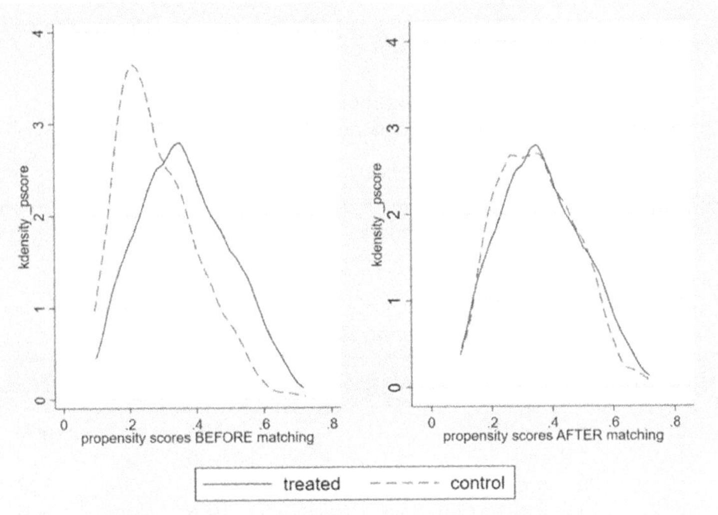

Figure 7.3 Propensity Score Overlapping between Treat and Control Groups, Before and After Matching for Improved Variety Seeds.

7.3.2.4   Impact of Integrated Soil Management on Yield and Income

Soil treatment is a vital CSA practice to maintain soil health. Overuse of fertilizers, pesticides, drought and other natural hazards degrade the fertile soil into hard soil. The soil becomes acidic and hard. Soil testing, recommended use of fertilizer doses, and soil treatment using gypsum, papermill sludges and other nutrients enhance the nutrition level of the soil. Soil treatment practices significantly positively impact yield and total agriculture income. ATT shows that the mean difference between adopters and non-adopters on net agricultural income is 27–34%. The adoption of soil treatment increases productivity by 2–3 quintals of paddy yield on average (Tables 7.8 and 7.9 and Figures 7.4 and 7.5).

Outcome variables	Matching methods	Improved variety seeds	
		ATT	*SE*
Yield per acre	Common radius calliper	1.25***	0.568
	Kernel	1.12*	0.578
	5-Nearest neighbour	0.646	0.662
Total agricultural income	Common radius calliper	0.259***	0.114
	Kernel	0.279***	0.116
	5-Nearest neighbour	0.291***	0.132

*Significant at 10% level; **Significant at 5% level; ***Significant at 1% level

Table 7.8 Average Treatment Effect for Yield and Total Agriculture Income

Outcome variables	Matching methods	Integrated soil management	
		ATT	*SE*
Yield per acre	Common radius calliper	2.85***	0.529
	Kernel	3.017***	0.537
	5-Nearest neighbour	2.59***	0.685
Total agricultural income	Common radius calliper	0.324***	0.1162
	Kernel	0.346***	0.1183
	5-Nearest neighbour	0.277**	0.1523

*Significant at 10% level; **Significant at 5% level; ***Significant at 1% level

Table 7.9 Impact of Covariates Balancing Before and After Matching

Adoption practice	Indicators of covariate balancing	Before matching	After matching
Crop rotation	Pseudo R^2	0.155	0.009
	LR chi^2	101.20	7.16
	p-value of log-likelihood	0.000	0.928
	Median absolute bias	19.6	5.1
	Total % bias reduction (%)	99.7	22.0

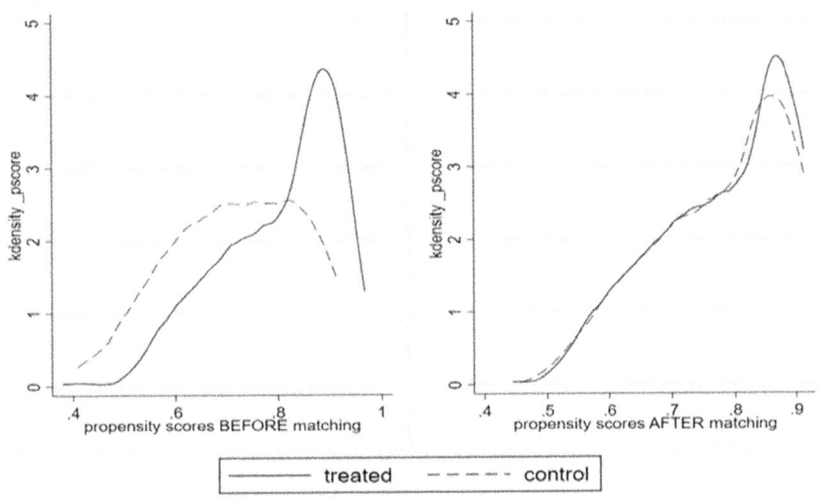

Figure 7.4 Propensity Score Overlapping between Treat and Control Groups, Before and After Matching for Soil Treatment.

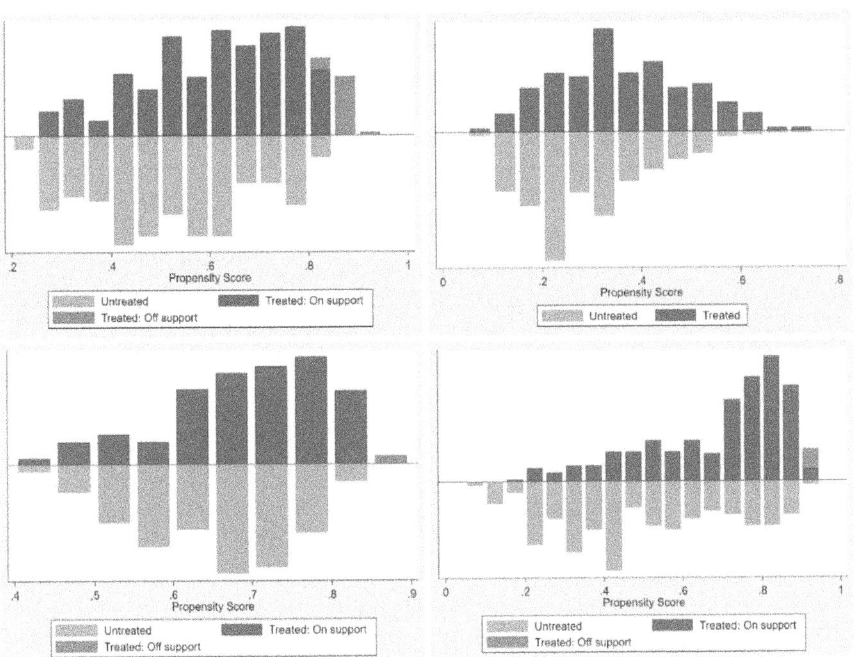

Figure 7.5 Common Support; Nearest Neighbour (5), Kernel, and Radius Matching Method.

7.4 Discussion

In general, the findings highlight the significance of innovations within CSAs at the farmer level in terms of constructing resistance to the effects of climate change and other issues on the farm that are related to production. Using CSA practices mitigates the negative effects of climate change on food production and farmers' incomes. When complementary CSA improvements such as improved variety seeds (stress-tolerant) and integrated agriculture practices (crop rotation and crop diversification) are used in diverse combinations, CSA's benefits increase to a greater extent. The increased impact may occur as a result of the combined effect of implementing the innovations, which improves soil nutrient usage, soil fertility, pest and disease stress, and climatic stress. The findings are consistent with the growing body of research that demonstrates how the adoption of multiple CSA practices can simultaneously have a multiplicative effect on a number of variables associated with welfare, including productivity and income. Certain studies – Makate et al. (2017), Teklewold et al. (2013), Khonje et al. (2018), Wainaina et al. (2018), Lunduka et al. (2019) – have also found a significant impact of CSA on the yield and income of farmers. Amadu et al. (2020) found that adopting CSA practices increased yield by 53% in the drought year of 2016. These studies discovered that smallholder farmers who used sustainable agricultural practices saw greater increases in both their productivity and their income. Additionally, it was observed that implementing many CA practices simultaneously had a cumulative effect on income. Expanding the usage of this technology would help farmers become more resilient to the adverse effects of climate change by increasing their household incomes and the likelihood that they will have strong food security.

Odisha is a vulnerable state with frequent climatic shocks that destroy the crops in the region multiple times. The adoption of CSA practices in this region and their cumulative benefit on productivity will motivate non-adopter farmers to adopt CSA practices. The farmers who are rich and have a consistently high income may not be in dire need of an extra INR 10,000–20,000 income, but for the marginal and small farmers, any increase in their income profile is desirable. If the income is higher, the farmers will come out of the vicious circle of poverty, and their lives will become more sustainable.

7.5 Conclusion and Policy Implication

The objective of this chapter is to analyse the impact of CSA adoption on farmers' income and yield per acre. We used our primary survey database of 494 households in Odisha. We have used two popular methods to find the impact of CSA on the productivity and income of the farmer households. In order to study the effect of adoption on crop yield and income, we had to ensure that no selection bias occurred. Therefore, the propensity score method was employed to ensure this. The results were as expected: there is a significant and positive impact of multiple CSA adoption on paddy yield and farm income. To remove self-selection bias and endogeneity, we have used two instruments, i.e., distance to the extension office and percentage of multiple adapters in a village. We analysed the impact using the TSLS method, which gave us an unbiased result and removed the endogeneity issue. The TSLS method shows that for every adoption of CSA, there is a positive increase in the percentage of income. All the results from the different models are similar, which implies that the instruments have worked correctly to predict the impact of the adoption of CSA practices. Much of the literature has justified the positive association of CSA with productivity and income, so this study is also in line with previous literature. Therefore, there is a clear indication of an increase in income and productivity associated with adopting CSA practices. The main challenge in the IV approach is identifying suitable instruments to confirm the exogeneity of the error term. As a result, it is important to interpret the estimates obtained from this approach with caution.

The finding suggests that there is a positive effect association between the adoption of CSA and increased farm income and productivity. These associations have various policy implications. The farmers' participation and peer interaction are needed to adopt the CSA practices. Thus, farmers should be educated about the benefits of participation, to enhance engagement and achieve policy goals. If farmer-to-farmer extension works effectively, it boosts the adoption and helps to increase income and productivity. The extension engagement should be more frequent, and the extension office setup should be located near the farmer so as to increase the adoption of CSA practices.

References

Ali, A., Hussain, I., Rahut, D. B. and Erenstein, O. (2018) 'Laser-land levelling adoption and its impact on water use, crop yields and household income: Empirical evidence from the rice-wheat system of Pakistan Punjab', *Food Policy*, 77, pp. 19–32.

Amadu, F. O., McNamara, P. E. and Miller, D. C. (2020) 'Understanding the adoption of climate-smart agriculture: A farm-level typology with empirical evidence from southern Malawi,' *World Development*, 126, p. 104692.

Branca, G., McCarthy, N., Lipper, L. and Jolejole, M. C. (2011) 'Climate-smart agriculture: A synthesis of empirical evidence of food security and mitigation benefits from improved cropland management', *Mitigation of Climate Change in Agriculture Series*, 3, pp. 1–42.

Challinor, A. J., Watson, J., Lobell, D. B., Howden, S. M., Smith, D. R. and Chhetri, N. (2014) 'A meta-analysis of crop yield under climate change and adaptation', *Nature Climate Change*, 4(4), pp. 287–291.

Choudhary, B. B., Dev, I., Singh, P., Singh, R., Sharma, P., Chand, K. and Kumar, N. (2022) 'Impact of soil and water conservation measures on farm productivity and income in the semi-arid tropics of Bundelkhand, Central India', *Environmental Conservation*, 49(4), pp. 263–271.

FAO. (2011) 'Adapt: Framework Programme on Climate Change Adaptation', 38. Retrieved from https://shorturl.at/nBCJO

Khatri-Chhetri, A., Aryal, J. P., Sapkota, T. B. and Khurana, R. (2016) 'Economic benefits of climate-smart agricultural practices to smallholder farmers in the Indo-Gangetic plains of India', *Current Science*, 110, 1251–1256. https://doi.org/10.18520/cs/v110/i7/1251-1256.

Khonje, M. G., Manda, J., Mkandawire, P., Tufa, A. H. and Alene, A. D. (2018) 'Adoption and welfare impacts of multiple agricultural technologies: Evidence from Eastern Zambia', *Agricultural Economics*, 49(5), pp. 599–609.

Kumar, P. and Sharma, P. K. (2020) 'Soil salinity and food security in India,' *Frontiers in Sustainable Food Systems*, 4, p. 533781.

Lan, L., Sain, G., Czaplicki, S., Guerten, N., Shikuku, K. M., Grosjean, G. and Läderach, P. (2018) 'Farm-level and community aggregate economic impacts of adopting climate smart agricultural practices in three mega environments', *PLoS One*, 13(11), p. e0207700.

Lipper, L., Thornton, P., Campbell, B. M., Baedeker, T., Braimoh, A., Bwalya, M. and Torquebiau, E. F. (2014) 'Climate-smart agriculture for food security', *Nature Climate Change*, 4(12), pp. 1068–1072.

Lopez-Ridaura, S., Frelat, R., van Wijk, M. T., Valbuena, D., Krupnik, T. J. and Jat, M. L. (2018) 'Climate smart agriculture, farm household typologies and food security: An ex-ante assessment from Eastern India', *Agricultural Systems*, 159, pp. 57–68.

Lunduka, R. W., Mateva, K. I., Magorokosho, C. and Manjeru, P. (2019) 'Impact of adoption of drought-tolerant maize varieties on total maize production in south Eastern Zimbabwe', *Climate and Development*, 11(1), pp. 35–46.

Makate, C., Makate, M. and Mango, N. (2018) 'Farm household typology and adoption of climate-smart agriculture practices in smallholder farming systems of southern Africa', *African Journal of Science, Technology, Innovation and Development. Taylor and Francis Ltd*, 10(4), pp. 421–439.

Makate, C., Makate, M. and Mango, N. (2017) 'Sustainable agriculture practices and livelihoods in pro-poor smallholder farming systems in southern Africa', *African Journal of Science, Technology, Innovation and Development*, 9(3), pp. 269–279.

Mishra, D., Sahu, N. C. and Sahoo, D. (2016) 'Impact of climate change on agricultural production of Odisha (India): A ricardian analysis', *"Regional Environmental Change*, 16(2), pp. 575–584.

Mohanty, B. B. and Lenka, P. K. (2019) 'Small farmers' suicide in Odisha', *Economic and Political Weekly*, 54(22), p. 51.

Mungai, L. M., Snapp, S., Messina, J. P., Chikowo, R., Smith, A., Anders, E., Richardson, R. B. and Li, G. (2016) 'Smallholder farms and the potential for sustainable intensification', *Front Plant Science*, 7, p. 1720.

Pretty, J. (2008) 'Agricultural sustainability: Concepts, principles and evidence', *Philosophical Transactions of the Royal Society B: Biological Sciences*, 363(1491), pp. 447–465.

Qureshi, M. R. N. M., Almuflih, A. S., Sharma, J., Tyagi, M., Singh, S. and Almakayeel, N. (2022) 'Assessment of the climate-smart agriculture interventions towards the avenues of sustainable production–Consumption', *Sustainability*, 14(14), p. 8410.

Sahu, N. C. and Mishra, D. (2013) 'Analysis of perception and adaptability strategies of the farmers to climate change in Odisha, India', *APCBEE Procedia*, 5, pp. 123–127.

Tanti, P. C., Jena, P. R., Aryal, J. P. and Rahut, D. B. (2022) 'Role of institutional factors in climate-smart technology adoption in agriculture: Evidence from an Eastern Indian state', *Environ Challenges*, 7, p. 100498.

Teklewold, H., Kassie, M., Shiferaw, B. and Köhlin, G. (2013) 'Cropping system diversification, conservation tillage and modern seed adoption in Ethiopia: Impacts on household income, agrochemical use and demand for labor', *Ecological Economics*, 93, p. 85.

The Hindu. (2022) India suffered income loss of $159 billion in key sectors due to extreme heat in 2021: Report. Retrieved from https://www.thehindu.com/business/Economy/india-suffered-income-loss-of-159-billion-in-key-sectors-due-to-extreme-heat-in-2021-report/article66035523.ece

Vermeulen, S. J., Aggarwal, P. K., Ainslie, A., Angelone, C., Campbell, B. M., Challinor, A. J. and Wollenberg, E. (2012) 'Options for support to agriculture and food security under climate change', *Environmental Science and Policy*, 15(1), pp. 136–144.

Wainaina, P., Tongruksawattana, S. and Qaim, M. (2018) 'Synergies between different types of agricultural technologies in the Kenyan small farm sector', *The Journal of Development Studies*, 54(11), pp. 1974–1990.

Wekesa, B. M., Ayuya, O. I. and Lagat, J. K. (2018) 'Effect of climate-smart agricultural practices on household food security in smallholder production systems: Micro-level evidence from Kenya', *Agriculture and Food Security*, 7(1), pp. 1–14.

Xiong, W., Van der Velde, M., Holman, I. P., Balkovic, J., Lin, E., Skalský, R., Porter, C., Jones, J., Khabarov, N. and Obersteiner, M. (2014) 'Can climate-smart agriculture reverse the recent slowing of rice yield growth in China?' *Agriculture, Ecosystems & Environment*, 196, pp. 125–136.

8 Forest Ecosystems in India and Linkage to Agricultural Sustainability

8.1 Introduction

India, known for its rich biodiversity and vast forest cover, plays a vital role in achieving sustainable agriculture. This chapter explores India's forest ecosystem and its significance in promoting agricultural sustainability. The study emphasizes the interdependence between forests and agriculture, highlighting the forest's contributions to sustainable farming. It examines how forests positively impact soil health, water management, climate regulation, and biodiversity conservation. The research analyses successful case studies and initiatives that integrate forest resources and traditional farming practices, showcasing innovative approaches to sustainable agriculture across India. These include agroforestry models, watershed management programmes, and community-led conservation efforts, highlighting the potential for forest-agriculture synergy and associated socioeconomic benefits. The chapter also discusses challenges and opportunities in harnessing India's forest ecosystem for sustainable agriculture, addressing land-use conflicts, deforestation, policy gaps, and the need for effective governance frameworks. The findings emphasize the importance of recognizing and incorporating forests into agricultural planning and policies. Collaborative efforts among farmers, local communities, policymakers, and researchers are crucial in developing strategies that utilize the multiple benefits of forests for sustainable agricultural development. Understanding and appreciating the integral role of India's forest ecosystem is vital for achieving food security, environmental conservation, and socioeconomic well-being. Embracing the concept of "Green Harmony", where forests and agriculture coexist and complement each other, India can pave the way towards a sustainable and resilient future.

With the projected global population reaching 9 billion by 2050 and an increasing shift towards urban areas, the demand for food security and ecosystem services becomes crucial. Forests play a direct and indirect role in meeting these needs, as approximately 25% of the global population depends on forests for subsistence, livelihoods, employment, and income generation (United Nations, 2017). However, the current scenario reveals alarming trends. Every year, 13 million hectares (ha) of

DOI: 10.4324/9781003290445-8

forests disappear, leading to adverse consequences for the planet Earth (FAO, 2020). Indigenous and economically disadvantaged communities often resort to unsustainable practices, exacerbating the problem (United Nations, 2021). Recognizing the importance of a healthy ecosystem, it becomes imperative to adopt sustainable practices and understand the changes within the ecosystem to address these challenges.

Conserving, sustainably managing, and restoring forests can contribute to multiple aspects of sustainable development, including economic growth, poverty alleviation, rule of law, food security, climate resilience, and biodiversity conservation (United Nations, 2021). Forest ecosystems have a positive association with sustainable development, as they support a range of economic activities. Subsistence agriculture and fisheries, in particular, provide livelihoods for rural communities. Protected and conserved areas are crucial for supporting the livelihoods of socially and economically vulnerable populations, ensuring the provision of ecosystem services, enhancing food security, mitigating pollution, and promoting human health (World Bank, 2020). Additionally, these areas play a significant role in providing alternative sources of clean and renewable energy, such as hydropower systems (MacKinnon et al., 2019).

The link between biodiversity and human health is increasingly recognized, with forests playing a vital role in regulating water flow, improving water quality, and serving as a source of medicinal resources (Marselle et al., 2021). Moreover, protected and conserved areas offer recreational opportunities that contribute to physical and mental well-being. Raising awareness about the importance of protecting biodiversity and creating protected and conserved areas is crucial for achieving sustainable development goals (SDGs). Equitable participation of women and marginalized groups in decision-making processes is essential, as is the fair sharing of benefits and costs. Urban planning that integrates biodiversity, such as the creation of urban parks, can contribute to more sustainable cities (Hernandez-Santin et al., 2023).

Unsustainable consumption patterns have detrimental impacts on ecosystems and biodiversity, emphasizing the need to preserve ecosystem functions and promote efficient resource use. Climate change further exacerbates the pressures on species and ecosystems, emphasizing the importance of protecting them to enhance climate change mitigation and adaptation (Olsson et al., 2019). Marine and terrestrial protected and conserved areas contribute directly to sustainable development by preserving marine resources, sustaining terrestrial habitats, and safeguarding against environmental crimes and conflicts over natural resources (Convention on Biological Diversity, 2016). Furthermore, the equitable governance of these areas, along with recognition of the rights of indigenous peoples and local communities, contributes to a better society.

In this context, India, renowned for its rich biodiversity and extensive forest cover, holds a significant key to addressing these challenges through sustainable agriculture practices. Past studies have emphasized the importance of forests in supporting sustainable agriculture. Forests provide essential ecosystem services, such as soil health improvement, water management, climate regulation, and biodiversity conservation (FAO, 2020). These services are vital for maintaining resilient and productive agricultural systems and ensuring long-term sustainability. India's forest ecosystem plays a crucial role in promoting sustainable agriculture by contributing to various aspects of farming. Forests enhance soil fertility through nutrient cycling, prevent erosion, and improve water retention capacity (Dubois, 2011). They also support pollination and natural pest control, reducing the reliance on chemical inputs (Dubois, 2011). Moreover, forests provide diverse habitats for beneficial organisms and help maintain biodiversity, which is essential for maintaining ecosystem resilience and agricultural productivity (Dubois, 2011).

The integration of forest resources and traditional farming practices has shown promising results in different regions of India. Agroforestry models, watershed management programmes, and community-led conservation efforts have demonstrated the potential for forest-agriculture synergy and associated socioeconomic benefits (Kearney et al., 2019; Telwala, 2023). These successful case studies showcase innovative approaches to sustainable agriculture, emphasizing the importance of harmonizing forest ecosystems with agricultural practices.

Despite the significant role of India's forest ecosystem in sustainable agriculture, challenges and opportunities exist. Land-use conflicts, deforestation, and policy gaps hinder the full realization of the potential benefits. Effective governance frameworks and robust conservation strategies are essential for promoting forest conservation and sustainable agricultural practices. Understanding the integral role of India's forest ecosystem in promoting sustainable agriculture is crucial for achieving food security, environmental conservation, and socioeconomic well-being. By embracing the concept of "Green Harmony", where forests and agriculture coexist and complement each other, India can pave the way towards a more sustainable and resilient future.

In this study, we aim to unveil the key role of India's forest ecosystem in sustainable agriculture. Through a comprehensive analysis of past research and case studies, we will highlight the positive impacts of forests on soil health, water management, climate regulation, and biodiversity conservation. Additionally, we will explore successful initiatives and innovative strategies that integrate forest resources and traditional farming practices. By shedding light on the challenges and opportunities, we will emphasize the importance of collaborative efforts between farmers, local communities, policymakers, and researchers in harnessing the full potential of India's forest ecosystem for sustainable agriculture.

8.2 Complementarity between Forest Ecosystem and Sustainable Agriculture: The Linkages

The complementarity between forest ecosystems and sustainable agriculture refers to the mutually beneficial relationship and synergies that can be fostered between these two domains. By recognizing and leveraging the interconnections and interdependencies between forests and agriculture, it is possible to promote practices that enhance both environmental conservation and agricultural productivity.

Biodiversity conservation: Forest ecosystems are vital repositories of biodiversity, housing a wide array of plant and animal species. By integrating forests into agricultural landscapes, it is possible to create ecological corridors and buffer zones that support diverse habitats and species. The presence of forests can contribute to the conservation of pollinators, natural predators of pests, and other beneficial organisms, thereby reducing the reliance on chemical inputs in agriculture.

Soil health and nutrient cycling: Forests play a crucial role in maintaining soil fertility and nutrient cycling. The dense root systems and organic matter in forests help to prevent soil erosion, improve soil structure, and enhance water infiltration. When forests are incorporated into agricultural systems through practices such as agroforestry, they can contribute to improved soil health, increased soil organic matter, and enhanced nutrient availability for crops, leading to higher agricultural productivity and resilience.

Water management: Forests act as natural water regulators, influencing the hydrological cycle by intercepting rainfall, promoting infiltration, and regulating streamflow. This can contribute to improved water availability for agricultural purposes, especially in regions prone to drought or water scarcity. Protecting and restoring forested watersheds can enhance water quality, reduce soil erosion, and maintain streamflow, benefiting both ecosystems and agricultural activities downstream.

Climate change mitigation and adaptation: Forests are vital in mitigating climate change as they sequester and store carbon dioxide through photosynthesis. By promoting afforestation, reforestation, and agroforestry practices, the agricultural sector can contribute to carbon sequestration, offsetting greenhouse gas emissions. Forests also provide shade, windbreaks, and microclimate regulation, which can enhance crop resilience to climate variability and extreme weather events.

Livelihoods and socioeconomic benefits: Forests offer a range of non-timber forest products, such as fruits, nuts, mushrooms, medicinal plants, and honey, which can diversify income sources for rural communities. Integrating sustainable forest management and agroforestry practices into agricultural systems can create additional income streams, promote rural livelihoods, and contribute to poverty alleviation.

Cultural and recreational values: Forests hold cultural and recreational significance, providing spaces for spiritual practices, traditional knowledge exchange, and recreation. By preserving forests and incorporating them into agricultural landscapes, communities can maintain their cultural heritage and promote ecotourism, generating economic opportunities while conserving natural resources.

Recognizing and fostering the complementarity between forest ecosystems and sustainable agriculture requires an integrated approach that considers ecological, economic, and social dimensions. It involves adopting practices such as agroforestry, watershed management, biodiversity conservation, and sustainable land-use planning. Policy frameworks, incentives, and capacity-building initiatives that promote sustainable agricultural practices while safeguarding forest ecosystems are essential for harnessing the synergies between the two domains and achieving long-term sustainability. The complementarity between forest ecosystems and sustainable agriculture has gained significant attention in the field of environmental science and agricultural sustainability.

8.2.1   Forest Ecosystem Services Supporting Sustainable Agriculture

Numerous studies have highlighted the ecosystem services provided by forests that directly contribute to sustainable agriculture (Perfecto et al., 2009; Jose, 2009; Milder et al., 2010; Kumar and Singh, 2020; Kumar et al., 2021). These services include soil fertility enhancement, water regulation, climate regulation, and biodiversity conservation. For instance, the presence of forests can improve soil health, reduce erosion, and enhance nutrient cycling, thereby promoting sustainable agricultural practices. Forests also play a crucial role in water regulation by acting as natural watersheds, ensuring a sustainable water supply for agricultural activities. Additionally, the diverse habitats found in forests support biodiversity, including pollinators and natural predators, which contribute to natural pest control in agricultural systems.

8.2.1.1   Agroforestry Systems as a Bridge between Forests and Agriculture

Agroforestry systems have emerged as a promising approach to harness the complementarity between forests and agriculture. These systems involve the intentional integration of trees within agricultural landscapes, combining the benefits of both systems. Studies have demonstrated that agroforestry practices improve soil fertility, increase carbon sequestration, enhance biodiversity, and provide additional sources of income for farmers (Lasco et al., 2014; Wartman et al., 2018; Sahoo et al., 2019). By promoting agroforestry, sustainable agricultural practices can be integrated with forest conservation, leading to more resilient and sustainable land use.

8.2.1.2   Sustainable Agricultural Practices Contributing to Forest Conservation

Sustainable agricultural practices also play a crucial role in the conservation and preservation of forest ecosystems. Research has shown that sustainable farming techniques, such as organic agriculture, integrated pest management, and conservation tillage, reduce the need for expansion into natural forests (Muhie, 2022). By minimizing deforestation and land degradation, sustainable agriculture practices help protect forest ecosystems and their associated biodiversity. Additionally, the use of environmentally friendly inputs and the adoption of agro-ecological approaches contribute to the long-term health and sustainability of forest ecosystems.

8.2.1.3 Policy Interventions and Collaborations

Government policies, international organizations, and local communities have recognized the importance of integrating forest conservation and sustainable agriculture. Various policy interventions, such as integrated land-use planning, sustainable forest management, and agro-ecological incentives, have been implemented to promote the complementarity between forests and agriculture. Collaborations between different stakeholders, including researchers, policymakers, and farmers, have also played a crucial role in advancing this agenda.

8.2.2 Complementarity between Forest Ecosystem and Sustainable Agriculture: The Literature Review

The literature examining complementarity between forest ecosystem and sustainable agriculture it categorized into two strands: Forest Ecosystem and Sustainable Agriculture, and Agroforestry and Sustainable Agriculture.

8.2.2.1 Forest Ecosystem and Sustainable Agriculture

The relationship between forest ecosystems and sustainable agriculture has gained significant attention in recent years. Forests provide a multitude of ecosystem services that are crucial for supporting sustainable agricultural systems (Perfecto et al., 2009; Milder et al., 2010). This literature review aims to provide an overview of the existing research on the interactions between forest ecosystems and sustainable agriculture, focusing on the benefits, challenges, and opportunities associated with their integration.

8.2.2.1.1 BIODIVERSITY AND HABITAT CONSERVATION

Forest ecosystems play a critical role in preserving biodiversity and providing habitats for numerous plant and animal species (Bhagwat et al., 2008; Perfecto et al., 2009). The literature demonstrates that incorporating forest elements, such as forest edges or forest fragments, into agricultural landscapes can promote biodiversity conservation and support the ecological balance necessary for sustainable agriculture (Tscharntke et al., 2012; Bommarco et al., 2013). For example, studies have shown that the presence of forest patches within agricultural landscapes enhances species richness and abundance of birds, bees, and other beneficial insects (Nicholls and Altieri, 2013; Miñarro and Prida, 2013).

8.2.2.1.2 NUTRIENT CYCLING AND SOIL FERTILITY

Forest ecosystems contribute to nutrient cycling and soil fertility, which are essential for sustainable agricultural production (Crews and Bender, 2011; van Noordwijk et al., 2014). The literature reveals that the presence of forests or agroforestry systems in agricultural landscapes can enhance soil organic matter, nutrient availability, and nutrient retention, leading to improved soil fertility and crop productivity (Nair, 2012; Datta et al., 2017). Research has shown that agroforestry practices, such as alley cropping and shade-grown coffee systems, improve nutrient cycling by increasing nutrient inputs from tree litter and root exudates, resulting in improved soil fertility (Jose, 2009; Kumar and Singh, 2020; Kumar et al., 2021).

8.2.2.1.3 WATER RESOURCE MANAGEMENT

Forest ecosystems play a crucial role in regulating water resources, including groundwater recharge, streamflow regulation, and water quality maintenance (Farley et al., 2005; Ellison et al., 2017). Research indicates that incorporating forests or agroforestry systems in agricultural landscapes can contribute to improved water management, reduce water runoff and erosion, and enhance water availability for agricultural purposes (Jackson et al., 2005; Zomer et al., 2016). For instance, studies have shown that agroforestry systems with tree cover reduce surface runoff, increase water infiltration, and improve water retention capacity compared to open-field agriculture (Dawson et al., 2015; Zhu et al., 2018; Hailu et al., 2000).

8.2.2.1.4 CLIMATE CHANGE MITIGATION AND ADAPTATION

Forests are important allies in climate change mitigation and adaptation efforts. They sequester carbon dioxide, help stabilize global climate patterns, and provide resilience to climate variability (Aba et al., 2017; Hisano et al., 2018). The literature suggests that integrating forest ecosystems or agroforestry practices in agricultural landscapes can contribute to carbon sequestration, reduce greenhouse gas emissions, and enhance the resilience of agricultural systems to climate change (Garrity et al., 2010; Sileshi et al., 2020). For example, studies have demonstrated that agroforestry systems with trees acting as windbreaks can reduce crop vulnerability to extreme weather events, such as strong winds and heatwaves, thereby enhancing climate resilience (Newaj et al., 2013; Chavan et al., 2014; Torralba et al., 2016).

8.2.2.1.5 SOCIO-ECONOMIC CONSIDERATIONS

The integration of forest ecosystems and sustainable agriculture also has socioeconomic implications. Research indicates that forest-based agricultural systems, such as agroforestry or forest gardens, can provide diverse livelihood opportunities, enhance food security, and support the well-being of local communities (Nair, 2012; Wartman et al., 2018; Sahoo et al., 2019). Additionally, the literature highlights the importance of involving local communities in decision-making processes and ensuring equitable benefits from the integration of forest ecosystems and agriculture (Newaj et al., 2013; Chavan et al., 2014).

The literature reviewed here emphasizes the multifaceted benefits of integrating forest ecosystems and sustainable agriculture. The inclusion of forest elements in agricultural landscapes contributes to biodiversity conservation, nutrient cycling, water resource management, climate change mitigation, and socioeconomic development. However, challenges such as land tenure issues, policy gaps, and knowledge gaps need to be addressed to facilitate the integration of forest ecosystems and agriculture. By leveraging the opportunities and addressing the challenges, the integration of forest ecosystems and sustainable agriculture can promote ecological resilience, support

8.2.2.2 Agroforestry and Sustainable Agriculture

Agroforestry is a land management system that combines the practices of agriculture and forestry in a mutually beneficial way. It involves integrating trees or shrubs with agricultural crops or livestock within the same area or on adjacent land. Agroforestry systems are designed to enhance ecological, economic, and social benefits by harnessing the interactions and synergies between trees, crops, and animals.

Agroforestry, the integration of trees and agriculture, has gained recognition as a sustainable land management practice that contributes to ecological resilience and supports sustainable agriculture (Jose, 2009; Nair, 2012; Albrecht et al., 2020; Rakotovao et al., 2022). Agroforestry systems have been shown to improve soil structure, nutrient cycling, and organic matter accumulation, resulting in enhanced soil fertility (Jose, 2009; Kumar et al., 2021). Trees in agroforestry systems play a vital role in carbon sequestration, mitigating climate change by sequestering atmospheric carbon dioxide (Nair, 2012; Kay et al., 2019). Furthermore, agroforestry practices have demonstrated their effectiveness in water conservation, reducing soil erosion, and promoting biodiversity conservation (Garrity et al., 2010; Kay et al., 2019).

Agroforestry systems have been found to provide additional sources of income and livelihood diversification for farmers through the production of various products, including timber, fruits, nuts, and non-timber forest products (Mbow et al., 2014; Kay et al., 2019). Studies have shown that agroforestry contributes to poverty reduction, food security, and improved resilience in rural communities (Mbow et al., 2014; FAO, 2020). Additionally, agroforestry systems foster social cohesion, support the preservation of traditional knowledge systems, and promote cultural practices (Tchoundjeu et al., 2010; Zomer et al., 2016).

Despite the numerous benefits, agroforestry adoption faces challenges such as limited access to quality planting material, technical knowledge, and financial resources (Garrity et al., 2010; Saha et al., 2021). Land tenure issues and policy gaps also hinder the widespread adoption of agroforestry (Kiptot and Franzel, 2012; Sileshi et al., 2020). Research highlights the need for capacity building, farmer training programmes, and policy interventions to address these barriers and facilitate agroforestry adoption (Garrity et al., 2010; Castle et al., 2021).

The literature identifies various opportunities for scaling up agroforestry systems. Strengthening research and extension services is crucial for knowledge dissemination and technology transfer to farmers (Zomer et al., 2016). Integrating agroforestry into national agricultural strategies and climate change mitigation/adaptation plans can provide policy support and attract investment in agroforestry practices (Torralba et al., 2016; Sileshi et al., 2020). Developing market linkages and value chains for agroforestry products can enhance economic incentives and market opportunities for farmers (Tchoundjeu et al., 2010; Asaah et al., 2011; Mbow et al., 2014).

Overall, agroforestry represents a holistic approach to land use that integrates trees with agricultural practices, providing a more sustainable and multifunctional system that supports both human livelihoods and the environment. Amidst the present climate-change scenario, agroforestry systems have gained significant recognition as valuable tools for climate change mitigation and ensuring food security (Makundi and Sathaye, 2004; Feliciano et al., 2018). Agroforestry systems offer a multitude of benefits, including sustainable production, diversification of household production, conservation of resources, groundwater recharge, and environmental enhancements. Agroforestry plays a vital role in mitigating climate change through the storage of carbon both above ground, in the form of biomass, and below ground, in soil carbon. Planting trees and grasses in degraded lands can enhance soil organic carbon levels (Kar et al., 2009), further contributing to climate-change mitigation through agroforestry practices. Agroforestry offers a comprehensive solution to various environmental challenges, including tillage, burning of leaf litter, deforestation, wildfires, and soil disturbance, by effectively storing carbon and minimizing emissions into the atmosphere. Agroforestry systems exhibit varying levels of carbon storage across different regions, with average estimates of approximately 9, 21, 50, and 63 Mg C ha^{-1} in semiarid, sub-humid, humid, and temperate regions, respectively (Jat et al., 2016). Globally, there is a significant potential to implement agroforestry systems on a land area ranging from 585 to 1,275 × 10^6 ha, capable of storing approximately 12–228 (with a median of 95) Mg C ha^{-1}, considering current climate and soil conditions (Dixon, 1995). In the context of India, Sathaye and Ravindranath (1998) projected a mitigation potential of 25.4 Mg C ha^{-1} for agroforestry systems, while in Pakistan, the estimated mitigation potential is 29.7 Mg C ha^{-1}.

In conclusion, agroforestry offers a range of environmental, social, and economic benefits, making it a promising approach for sustainable agriculture. Addressing challenges related to access to resources, policy gaps, and technical knowledge is crucial for the widespread adoption of agroforestry. Strengthening research, policy support, and market development are important strategies to leverage the opportunities and overcome barriers, ultimately promoting the adoption of agroforestry as a pivotal contributor to sustainable agriculture and the well-being of farmers and ecosystems.

The literature review highlights the significant research conducted on the complementarity between forest ecosystems and sustainable agriculture. It underscores the importance of forest ecosystem services in supporting sustainable agricultural practices and the reciprocal benefits that sustainable agriculture brings to forest conservation. Agroforestry systems have emerged as a promising approach to integrate both systems and promote sustainable land use. However, further research is needed to address gaps in knowledge and to assess the long-term impacts of these practices on both forest ecosystems and agricultural sustainability. By fostering collaboration, policy interventions, and continued research, the complementarity between forest ecosystems and sustainable agriculture can be effectively harnessed to achieve environmental sustainability and food security.

8.3   Preservation of the Indian Forest Ecosystem: Assessing Conservation Efforts

The world has approximately 4.06 billion ha of forests, accounting for about 31% of the total land area (FAO, 2020). Forest distribution is uneven across regions and populations (FAO, 2020). The tropical region holds the largest share at 45%, followed by the boreal domain at 21%, and the temperate and subtropical domains at 16% and 11%, respectively. Five countries, including Russia (20%), Brazil (12%), Canada (9%), the United States (8%), and China (5%), possess more than half (54%) of the world's forests (FAO, 2020). Since 1990, there has been a significant loss of 178 million ha of forest, with a decreasing trend over time (FAO, 2020). Africa and South America experienced the highest rates of forest loss, while Asia observed the highest net gain of forest area in 2010–2020 (FAO, 2020).

The most recent decade saw a deceleration in the rate of decline of net forest loss, primarily attributed to a reduction in the expansion of forests. In terms of specific regions, Africa experienced the highest annual rate of net forest loss during 2010–2020, amounting to 3.9 million ha, closely followed by South America with 2.6 million ha. In Africa, the rate of net forest loss has witnessed a consistent increase over each of the three decades since 1990. Conversely, in South America, there has been a significant decline, with the rate in 2010–2020 approximately half of what it was during 2000–2010. On the other hand, Asia experienced the highest net gain of forest area in 2010–2020, followed by Oceania and Europe. However, both Europe and Asia observed a significant decrease in the rates of net gain in 2010–2020 compared to 2000–2010. Oceania, on the other hand, experienced net losses of forest area during the periods of 1990–2000 and 2000–2010. It is estimated that a staggering 420 million ha of forest have been lost globally due to deforestation since 1990. Fortunately, there has been a substantial decline in the rate of forest loss. In the most recent five-year period (2015–2020), the annual rate of deforestation was estimated at 10 million ha, down from 12 million ha during

2010–2015. As the objective of the study is related to the Indian context of forest ecosystem and sustainable agriculture, we presented the details below.

Table 8.1 presents the forest cover in India for various years, starting from 1987 to 2015. The forest cover is measured in square kilometres (sq. km). Let's analyse the trends and changes in forest cover during this period:

1987: The forest cover was recorded as 640,819 sq. km.

1989 to 1995: The forest cover remained relatively stable, with slight fluctuations between 638,804 and 639,386 sq. km.

1997: There was a noticeable decrease in forest cover, with the area dropping to 633,397 sq. km.

1999: The forest cover started to recover slightly, reaching 637,293 sq. km.

2001: A significant increase in forest cover was observed, with the area expanding to 653,898 sq. km.

2003: The upwards trend continued, and the forest cover reached 677,816 sq. km.

2005: The forest cover experienced further growth, reaching 690,171 sq. km. This marked a significant increase compared to the previous years.

2009–2011: The forest cover remained relatively stable, with minor fluctuations between 690,899 and 692,394 sq. km.

2013: There was a notable increase in forest cover, with the area expanding to 697,898 sq. km.

2015: The forest cover continued to rise, reaching 701,673 sq. km.

Overall, the forest cover in India showed fluctuations and variations during the given period. After a relatively stable phase in the late 1980s and early 1990s, there was a slight decline in forest cover in 1997, followed by a gradual

Table 8.1 Forest Cover (in sq. km.) in India, 1987–2015

Year	Forest cover (in sq. km.)
1987	640,819
1989	638,804
1991	639,364
1993	639,386
1995	638,879
1997	633,397
1999	637,293
2001	653,898
2003	677,816
2005	690,171
2009	692,394
2011	690,899
2013	697,898
2015	701,673

Source: Indiastat.com.

recovery. From 2001 to 2015, there was a generally positive trend, with increasing forest cover observed in most years.

It is important to note that while these numbers indicate changes in forest cover, they do not provide information about the quality, composition, or biodiversity of the forests. Additionally, the rate of deforestation, reforestation, and forest degradation cannot be determined solely from this data. A comprehensive analysis considering additional factors would be necessary to evaluate the overall state and trends of forests in India.

Table 8.2 provided shows the class-wise forest cover in India for the years 2004, 2008–2009, 2010–2011, 2013–2014, 2015–2016, and 2017–2018. The forest cover is measured in sq. km, and the percentages indicate the proportion of forest cover relative to the total geographic area of India. Let's analyse the forest cover data for each class over the mentioned years:

Very Dense Forest:
In 2004, the area covered by very dense forests was 54,569 sq. km, accounting for 1.66% of the total geographic area. From 2008–2009 to 2017–2018, there was a gradual increase in the area of very dense forests, reaching 99,278 sq. km (3.02%) in 2017–2018. Overall, there was an increase in the area of very dense forests during the given time period.

Moderately Dense Forest:
In 2004, the area covered by moderately dense forests was 332,647 sq. km, constituting 10.12% of the total geographic area. From 2008–2009 to 2017–2018, there was a slight decrease in the area of moderately dense forests, stabilizing around 308,000 sq. km (9.38%). Overall, the moderately dense forest cover remained relatively stable with a slight decline over the years.

Open Forest:
In 2004, the area covered by open forests was 289,872 sq. km, representing 8.82% of the total geographic area. From 2008–2009 to 2017–2018, there was a fluctuation in the area of open forests, but it remained relatively constant, ranging from 295,000 sq. km (8.99%) to 304,000 sq. km (9.26%). Overall, the open forest cover did not show a significant change during the given time period.

Total Forest Cover:
The total forest cover includes the area of very dense, moderately dense, and open forests combined. In 2004, the total forest cover in India was 690,171 sq. km, accounting for 20.6% of the total geographic area. From 2008–2009 to 2017–2018, there was a slight increase in the total forest cover, reaching 712,249 sq. km (21.67%) in 2017–2018. Overall, the total forest cover showed a modest increase over the years.

Scrub:
The area covered by scrublands, which are areas with low, stunted vegetation, was 38,475 sq. km (1.17%) in 2004. From 2008–2009 to 2017–2018, there was a gradual increase in the area of scrublands, reaching 46,297 sq. km (1.41%) in 2017–2018. Overall, the scrubland cover experienced a gradual expansion during the given time period.

Table 8.2 Class-wise Forest Cover in India (2004, 2008–2009, 2010–2011, 2013–2014, 2015–2016, and 2017–2018) (Area in sq. km).

Class	2004		2008–2009		2010–2011		2013–2014		2015–2016		2017–2018	
	Area	% Age of geographic Area	Area	% Age of geographic Area	Area	% Age of geographic Area	Area	% Age of geographic Area	Area	% Age of geographic Area	Area	% Age of geographic Area
Forest cover												
Very dense forest	54,569	1.66	83,471	2.54	83,502	2.54	85,904	2.61	98,158	2.99	99,278	3.02
Moderately dense forest	332,647	10.12	320,736	9.76	318,745	9.7	315,374	9.59	308,318	9.38	308,472	9.38
Open forest	289,872	8.82	287,820	8.75	295,651	8.99	300,395	9.14	301,797	9.18	304,499	9.26
Total forest cover*	690,171	20.6	692,027	21.05	697,898	21.23	701,495	21.34	708,273	21.54	712,249	21.67
Scrub	38,475	1.17	42,176	1.28	41,383	1.26	41,362	1.26	45,979	1.4	46,297	1.41
Non-forest	2,571,700	78.23	2,553,060	77.67	2,547,982	77.51	2,544,228	77.4	2,533,217	77.06	2,528,923	76.93
Total geographic area	3,287,263	100	3,287,263	100	3,287,263	100	3,287,263	100	3,287,469	100	3,287,469	100

Source: Indiastat.com.

Non-forest:

The non-forest area, which includes areas not classified as forests or scrub, was significantly larger than the forest cover in each year. In 2004, the non-forest area was 2,571,700 sq. km, accounting for 78.23% of the geographic area. The non-forest area decreased slightly over the years and reached 2,528,923 sq. km in 2017–2018, representing 76.93% of the geographic area. Overall, the forest cover in India, including very dense, moderately dense, open, and scrub forests, showed a slight increase from 2004 to 2017–2018. However, non-forest areas still dominate the geographic landscape, indicating the need for continued conservation efforts and sustainable forest management practices.

The results reveal substantial changes in carbon stocks in different components of Indian forest land between 1994 and 2004 (Table 8.3). The key findings are as follows:

Above Ground Biomass:

In 1994, the carbon stock in above-ground biomass was 1,784 million tonnes, which increased to 2,101 million tonnes in 2004. The net change in carbon stock for above-ground biomass was 317 million tonnes, indicating a significant increase over the decade.

Below Ground Biomass:

Carbon stock in below-ground biomass showed an increase from 563 million tonnes in 1994 to 663 million tonnes in 2004. The net change in carbon stock for below-ground biomass was 100 million tonnes, highlighting a notable growth during the study period.

Dead Wood:

The carbon stock in dead wood increased from 19 million tonnes in 1994 to 25 million tonnes in 2004. The net change in carbon stock for dead wood was 6 million tonnes, indicating a moderate increase over the decade.

Litter:

Carbon stock in litter rose from 104 million tonnes in 1994 to 121 million tonnes in 2004. The net change in carbon stock for litter was 17 million tonnes, suggesting a noticeable increase during the study period.

Table 8.3 Carbon Stock in Forest Land by Various Carbon Containing Components in India (1994 and 2004) (in Million Tonne)

Component	1994	2004	Net change in carbon stock
Above ground biomass	1,784	2,101	317
Below ground biomass	563	663	100
Dead wood	19	25	6
Litter	104	121	17
Soil	3,601	3,753	152
Total	6,071	6,663	592

Source: Indiastat.com.

Soil:

Carbon stock in soil showed an increase from 3,601 million tonnes in 1994 to 3,753 million tonnes in 2004. The net change in carbon stock for soil was 152 million tonnes, signifying a substantial growth during the study period.

The study provides valuable insights into the changes in carbon stock in forest land in India between 1994 and 2004. The significant increases observed in above-ground biomass, below-ground biomass, dead wood, litter, and soil carbon stocks indicate the potential of Indian forests in sequestering and storing carbon. These findings underscore the importance of implementing effective forest management and conservation strategies to preserve and enhance carbon stocks, contributing to global efforts in climate change mitigation. Further research and continued monitoring are crucial to track long-term trends in carbon stocks and inform sustainable forest management practices.

The provided data represents the area afforested and the cumulative afforestation expenditure in India from 1951–1956 to 2015–2016 (Table 8.4). The figures are given in ha (1 sq. km equals 100 ha) for the area afforested and in crores of rupees for the expenditure. Let's analyse the data in detail.

During the first plan period from 1951 to 1956, a total of 52,000 ha were afforested, with an expenditure of 1.28 crores. This was the starting point for tracking afforestation efforts in India. In the subsequent plan periods, the area afforested and the cumulative afforestation increased significantly. For example, in the second plan period (1956–1961), 311,000 ha were afforested, resulting in a cumulative afforestation of 363,000 ha. The expenditure during this period was 6.86 crores, bringing the cumulative expenditure to 8.14 crores. The trend continued, with each plan period showing an increase in afforestation. By the sixth plan period (1980–1985), a substantial area of 4,650,000 ha was afforested, contributing to a cumulative afforestation of 8,206,000 ha. The expenditure during this period was 926.01 crores, and the cumulative expenditure reached 1,167.02 crores. From 1985 to 1997, there was a significant boost in afforestation efforts, with the area afforested reaching 8,863,000 ha in the fifth plan period (1985–1990) and 7,950,000 ha in the sixth plan period (1992–1997). The cumulative afforestation during this period rose to 28,131,000 ha.

During the seventh plan period (1997–2002), the area afforested increased to 8,050,000 ha, leading to a cumulative afforestation of 36,181,000 ha. The expenditure during this period was significantly higher at 7,350.5 crores, resulting in a cumulative expenditure of 15,964.06 crores. In the subsequent years, the rate of afforestation fluctuated, with varying areas afforested and expenditures incurred. However, there was a general upwards trend in both parameters. By the last recorded period (2015–2016), a total of 36,245,000 ha were afforested, with a cumulative afforestation of 38,245,000 ha. The expenditure during this period was 94.16 crores, bringing the cumulative expenditure to 19,555.83 crores. The data indicates a consistent effort by the Indian government to promote afforestation and allocate funds for this purpose. The increasing cumulative afforestation figures demonstrate the progress made in expanding forest cover in the country over the years. The corresponding cumulative expenditure highlights the financial resources invested in afforestation initiatives.

Table 8.4 Area and Expenditure on Afforestation in India (1951–1956 to 2015–2016) (1 sq. km = 100 ha).

Plan period	Area afforested in plan period (in 000 ha)	Cumulative (in 000 ha)	Afforestation expenditure in plan period (Rs. in crore)	Cumulative (Rs. in crore)
1951–1956	52	52	1.28	1.28
1956–1961	311	363	6.86	8.14
1961–1966	583	946	21.13	29.27
1966–1969	453	1,399	23.02	52.29
1969–1774	714	2,113	44.34	96.63
1974–1779	1,221	3,334	107.28	203.91
1979–1980	222	3,556	37.1	241.01
1980–1985	4,650	8,206	926.01	1,167.02
1985–1990	8,863	17,069	2,426.63	3,593.65
1990–1991	1,387	18,456	627.79	4,221.44
1991–1992	1,725	20,181	705.72	4,927.16
1992–1997	7,950	28,131	3,686.4	8,613.56
1997–2002	8,050	36,181	7,350.5*	15,964.06
2002–2003	405	36,586	151.26	16,115.32
2003–2004	283	36,869	207.98	16,323.3
2004–2005	107	36,976	233	16,556.3
2005–2006	54	37,030	248.12	16,804.42
2006–2007	0	37,030	292.75	17,097.17
2007–2008	493	37,523	392.95	17,490.12
2008–2009	173	37,696	345.62	17,835.74
2009–2010	104	37,800	318.17	18,153.91
2010–2011	57	37,857	309.99	18,463.9
2011–2012	141	37,998	303	18,766.9
2012–2013	56	38,054	193.37	18,960.27
2013–2014	81	38,135	257.62	19,217.89
2014–2015	74	38,209	243.78	19,461.67
2015–2016	36	38,245	94.16	19,555.83

Source: Indiastat.com.

It is important to note that afforestation efforts are crucial for ecological balance, biodiversity conservation, and mitigating the adverse effects of deforestation and climate change. The data reflects India's commitment to environmental sustainability and the recognition of the importance of forests in maintaining a healthy ecosystem.

8.4   The Legislation on Forest Conservation and Forest Ecosystem

The literature on the interaction of natural protected areas with local communities is considerable. Since the 1980s, research regarding environmental issues emphasizes the importance of continuous dialogue conservation of forest in order to achieve sustainability goals. There are several forest legislation act implemented by the government in order to preserve a sustainable environment; those are listed in Table 8.5.

Table 8.5 The Legislation on Forest Conservation and Forest Ecosystem

Legislation	Year of legislation	Brief description
Easement Act	1882	Under this law, private individuals are granted the rights to utilize groundwater as it is considered an inherent part of the land. Simultaneously, the law stipulates that all surface water is the property of the state, thereby establishing it as a state-owned resource.
Indian Fisheries Act	1897	This law sets forth two categories of penal offences, enabling the government to take legal action against individuals who employ dynamite or other explosive substances for the purpose of catching or exterminating fish, including poisonous fish, whether in coastal or inland areas, with the intention to cause harm or death.
Factories Act	1948 and amended in 1987	It was the pioneer in expressing concern for the working conditions of labourers. The 1987 amendment has further enhanced its environmental emphasis and broadened its scope to include hazardous processes.
River Boards Act	1956	Under this law, the states are empowered to collaborate with the central government in establishing an Advisory River Board, which serves as a mechanism to address and resolve issues pertaining to inter-state cooperation.
Merchant Shipping Act	1970	The primary objective of this initiative is to effectively manage and address waste generated by ships within a designated radius along coastal areas.
Wildlife Protection Act	1972–1973 and amendment in 1991	This legislation encompasses comprehensive provisions for safeguarding the well-being of birds and animals, including all aspects related to their habitat, water sources, and the forests that serve as their sustenance.
Water (Prevention and Control of Pollution) Act	1974	This law establishes a comprehensive institutional framework aimed at the prevention and mitigation of water pollution. It sets forth guidelines for maintaining water quality and regulating the discharge of effluents. Industries that produce pollutants are required to obtain permission before releasing waste into water bodies designated for effluent disposal. Under this act, the formation of the Central Pollution Control Board (CPCB) was mandated.

(*Continued*)

Table 8.5 (Continued)

Legislation	Year of legislation	Brief description
Water (Prevention and Control of Pollution) Cess Act	1977	This legislation enables the imposition and collection of cess or fees on industries and local authorities that utilize water resources.
Water (Prevention and Control of Pollution) Cess Rules	1978	This law encompasses standardized definitions and specifies the type and placement of metres that all water consumers are obligated to install.
Forest (Conservation) Act and Rules	1980–1981	This law aims to ensure the protection and conservation of forests.
Air (Prevention and Control of Pollution) Act	1981	This legislation is designed to regulate and mitigate air pollution. It assigns the responsibility of enforcing the provisions of this act to the Central Pollution Control Board (CPCB).
Air (Prevention and Control of Pollution) Rules	1982	This legislation outlines the protocols governing board meetings and specifies the powers vested in them.
Environment (Protection) Act	1986	This law grants the central government the authority to safeguard and enhance environmental quality, regulate and minimize pollution from various origins, and prohibit or impose limitations on the establishment and operation of industrial facilities based on environmental considerations.
Environment (Protection) Rules	1986	It establishes procedural guidelines for determining the standards governing the emission or discharge of environmental pollutants.
Air (Prevention and Control of Pollution) Amendment Act	1987	This legislation grants authority to both the central and state pollution control boards to effectively address severe air pollution emergencies.
Objective of Hazardous Waste (Management and Handling) Rules	1989	To regulate the generation, collection, treatment, importation, storage, and handling of hazardous waste.
Manufacture, Storage, and Import of Hazardous Rules	1989	It provides definitions for the relevant terminology and establishes an authority responsible for conducting annual inspections of industrial activities involving hazardous chemicals and isolated storage facilities.
Manufacture, Use, Import, Export, and Storage of Hazardous Micro-organisms/ Genetically Engineered Organisms or Cells Rules	1989	It was introduced with the objective of safeguarding the environment, nature, and public health concerning the application of gene technology and microorganisms.

(Continued)

Table 8.5 (Continued)

Legislation	Year of legislation	Brief description
Coastal Regulation Zone Notification	1991	This legislation imposes regulations on a range of activities, including construction, to ensure proper oversight. Additionally, it provides a degree of protection for backwaters and estuaries.
Public Liability Insurance Act and Rules	1991 and amendment in 1992	It was formulated to establish provisions for public liability insurance with the aim of promptly providing relief to individuals affected by accidents occurring during the handling of hazardous substances.
National Environmental Tribunal Act	1995	It has been established to grant compensation for damages caused to individuals, property, and the environment resulting from any activity involving hazardous substances.
National Environment Appellate Authority Act	1997	It has been established to handle appeals concerning the limitations imposed on areas where specific industries or activities are conducted, as prescribed under the EPA, while ensuring the implementation of certain safeguards.
Biomedical waste (Management and Handling) Rules	1998	It is a legally binding requirement for healthcare institutions to establish streamlined procedures for the appropriate handling of hospital waste, including segregation, disposal, collection, and treatment.
Environment (Siting for Industrial Projects) Rules	1999	It establishes comprehensive provisions regarding restricted zones for industrial siting, precautionary measures for site selection, and the necessary aspects of environmental protection that should be incorporated during the implementation of industrial development projects.
Municipal Solid Wastes (Management and Handling) Rules	2000	This legislation is applicable to all municipal authorities entrusted with the tasks of collecting, segregating, storing, transporting, processing, and disposing of municipal solid waste.
Ozone Depleting Substances (Regulation and Control) Rules	2000	This has been established to regulate the production and consumption of ozone-depleting substances.

(Continued)

Table 8.5 (Continued)

Legislation	Year of legislation	Brief description
Batteries (Management and Handling) Rules	2001	This legislation applies to all manufacturers, importers, re-conditioners, assemblers, dealers, auctioneers, consumers, and bulk consumers engaged in the manufacturing, processing, sale, purchase, and usage of batteries or components. Its purpose is to regulate and ensure the environmentally sound disposal of used batteries.
Noise Pollution (Regulation and Control) (Amendment) Rules	2002	It establishes the necessary terms and conditions to mitigate noise pollution, allowing the use of loudspeakers or public address systems during night hours (from 10:00 p.m. to 12:00 midnight) for cultural or religious festivities, subject to specific regulations.
Biological Diversity Act	2002	It is an act designed to ensure the conservation of biological diversity, promote the sustainable utilization of its components, and establish fair and equitable sharing of the benefits derived from the utilization of biological resources and the knowledge associated with them.
The National Green Tribunal Act	2010	The National Green Tribunal Act, 2010 (No. 19 of 2010) (NGT Act) was enacted with the primary objectives of establishing an NGT for the efficient and prompt resolution of cases pertaining to environmental protection, conservation of forests, and other natural resources. The NGT Act also aims to enforce legal rights related to the environment, provide relief and compensation for damages to individuals and property, and address associated matters.
Coastal Regulation Zone Notification	2011	The Ministry of Environment and Forests released the Coastal Regulation Zone Notification (Notification no. S O. 19(E), dated 6 January 2011) with the aim of safeguarding the livelihoods of fishing communities and local residents in coastal areas. The notification seeks to conserve and protect coastal stretches, promote sustainable development guided by scientific principles, and address the risks posed by natural hazards and rising sea levels caused by global warming.

(Continued)

Table 8.5 (Continued)

Legislation	Year of legislation	Brief description
Wildlife (Protection) Amendment Act	2013	The Bill aims to amend the Wild Life (Protection) Act, 1972, which is in place to safeguard and preserve wild animals, birds, and plants. The Act encompasses the management of their habitats and the regulation and control of trade or commerce associated with wildlife.
Solid Waste Management Rules	2016	They emphasize the segregation of waste at its source, mandate the responsibility of manufacturers to dispose of sanitary and packaging waste, and introduce user fees for collection, disposal, and processing of waste from bulk generators. Furthermore, it is recommended that biodegradable waste should be processed, treated, and composted or converted into biogas on-site whenever feasible, with any remaining waste to be handed over to waste collectors or agencies as directed by local authorities. These rules also encourage the use of compost, conversion of waste into energy, and the revision of criteria for landfill locations and capacities.
Wetlands (Conservation and Management) Rules	2017	It seeks to foster the integrated and sustainable management of coastal and marine areas in India, with a focus on enhancing the welfare and prosperity of traditional coastal and island communities. Additionally, it aims to advance the sustainability of coastal regions by fostering partnerships, promoting conservation practices, conducting scientific research, and harnessing knowledge for the benefit and well-being of present and future generations.
Electronic Waste Management	2018	The amendment in rules aims to streamline the management of E-waste generated within the country by directing it towards authorized dismantlers and recyclers, thereby formalizing the e-waste recycling sector. The revised rules highlight the need for collection targets under the Extended Producer Responsibility (EPR) provision, which have been updated to include targets for newly established producers who have recently commenced their sales operations

Source: Online and books

In India, a range of legislative measures has been implemented between 2011 and 2021 to effectively address the pressing issues of forest conservation and the protection of forest ecosystems. These measures have played a crucial role in safeguarding the country's valuable forest resources. Here are some key legislations enacted during this period.

The Compensatory Afforestation Fund Act (2016) stands out as a significant legislation. It establishes the National Compensatory Afforestation Fund and State Compensatory Afforestation Fund, serving as dedicated funds to manage and utilize resources received for compensatory afforestation, additional afforestation, and ecological restoration efforts. These funds play a crucial role in promoting sustainable afforestation and restoring ecosystems in areas where forests have been diverted for non-forest purposes.

The Forest Rights Act (2006), although enacted prior to 2011, continued to be implemented and amended during this period. This landmark legislation recognizes and vests forest rights and occupation in forestland to forest-dwelling communities and tribal groups. The act aims to empower these communities, enabling them to participate in decision-making processes related to forest management and conservation. By recognizing the traditional rights of forest-dwelling communities, this act promotes their active involvement in sustainable forest practices.

The National Green Tribunal Act (2010) is another notable legislation introduced during this period. It establishes the National Green Tribunal (NGT) as a specialized judicial body responsible for the expeditious resolution of cases pertaining to environmental protection, including those concerning forests and forest ecosystems. The NGT plays a vital role in ensuring the effective enforcement of environmental laws and regulations, thereby contributing to the conservation of forests and biodiversity.

The Wildlife (Protection) Amendment Act (2013) is an important amendment to the Wildlife (Protection) Act of 1972. It enhances penalties for offences related to wildlife conservation, including illegal hunting, poaching, and trafficking of wildlife and their products. The amendment reflects the government's commitment to strengthening wildlife conservation efforts and combatting illegal activities that threaten the rich biodiversity of India's forests.

In 2018, the Compensatory Afforestation Fund Management and Planning Authority (CAMPA) Guidelines were introduced. These guidelines lay down the procedures and mechanisms for the efficient utilization and management of funds under the Compensatory Afforestation Fund Act. The focus is primarily on activities such as afforestation, regeneration, and forest conservation. The guidelines ensure that the funds are utilized effectively to restore and protect forest ecosystems, promoting sustainable practices across the country.

The Indian Forest (Amendment) Act (2017) is a notable legislation aimed at promoting bamboo cultivation in non-forest areas. The amendment recognizes bamboo as a grass and removes restrictions on its transit and use.

This change in classification facilitates the commercial cultivation of bamboo, which supports livelihoods and economic growth while contributing to the conservation of forests.

While not specifically targeting forests, the Environmental Impact Assessment (EIA) Notification (2020) is an essential regulation governing the environmental clearance process for various projects, including those with potential impacts on forest ecosystems. The notification emphasizes the assessment of environmental and ecological impacts, including those on forests, as a crucial part of project approval processes. By incorporating forest ecosystem considerations, this notification ensures that projects are developed sustainably, minimizing adverse impacts on forests and their biodiversity.

These legislations represent significant strides taken by India in the realm of forest conservation and forest ecosystem management between 2011 and 2021. It is important to note that there may also be additional regional or state-specific laws and regulations that further contribute to forest conservation efforts. For comprehensive and up-to-date information on specific legislation, it is advisable to refer to official government sources and legal documents.

8.5 Conclusion

This chapter sheds light on the paramount importance of India's forest ecosystem in fostering sustainable agriculture and unveils the concept of "Green Harmony". India, renowned for its rich biodiversity and vast forest cover, possesses a vital asset that can contribute significantly to achieving agricultural sustainability. The interdependence between forests and agriculture is emphasized, underscoring the forest's indispensable contributions to sustainable farming practices.

Through a comprehensive analysis, this study highlights the myriad benefits of forests in promoting agricultural sustainability. Forests positively impact critical factors such as soil health, water management, climate regulation, and biodiversity conservation, all of which are essential for maintaining resilient and productive agricultural systems. The integration of forest resources and traditional farming practices is explored through successful case studies and innovative initiatives across India. Agroforestry models, watershed management programmes, and community-led conservation efforts are among the pioneering approaches discussed, illustrating the potential for synergistic relationships between forests and agriculture. These initiatives not only enhance agricultural productivity but also offer significant socioeconomic advantages, benefiting local communities and fostering sustainable development.

However, challenges such as land-use conflicts, deforestation, and policy gaps must be addressed to fully harness India's forest ecosystem for sustainable agriculture. Effective governance frameworks, guided by collaboration among farmers, local communities, policymakers, and researchers, are essential in overcoming these challenges and promoting the integration of forests into agricultural planning and policies.

Embracing the concept of "Green Harmony" holds immense promise for India's sustainable and resilient future. By recognizing and appreciating the integral role of forests in achieving food security, environmental conservation, and socioeconomic well-being, India can set a powerful example for the world. The successful implementation of sustainable practices that coexist and complement both forests and agriculture will pave the way for a harmonious and balanced future, where the nation's natural resources are wisely managed and the benefits are equitably distributed.

In conclusion, India's forest ecosystem serves as a valuable ally in the pursuit of sustainable agriculture. By embracing the principles of "Green Harmony", India can unlock the full potential of its forests, leading the way towards a future where agriculture and nature thrive in a mutually beneficial relationship, ensuring a sustainable, resilient, and prosperous society for generations to come.

References

Aba, S. C., Ndukwe, O. O., Amu, C. J. and Baiyeri, K. P. (2017) 'The role of trees and plantation agriculture in mitigating global climate change', *African Journal of Food, Agriculture, Nutrition and Development*, 17(4), pp. 12691–12707.

Albrecht, M., Kleijn, D., Williams, N. M., Tschumi, M., Blaauw, B. R., Bommarco, R., … and Sutter, L. (2020) 'The effectiveness of flower strips and hedgerows on pest control, pollination services and crop yield: A quantitative synthesis,' *Ecology Letters*, 23(10), pp. 1488–1498.

Asaah, E. K., Tchoundjeu, Z., Leakey, R. R., Takousting, B., Njong, J. and Edang, I. (2011) 'Trees, agroforestry and multifunctional agriculture in Cameroon', *International Journal of Agricultural Sustainability*, 9(1), pp. 110–119.

Bhagwat, S. A., Willis, K. J., Birks, H. J. and Whittaker, R. J. (2008) 'Agroforestry: A refuge for tropical biodiversity?', *Trends in Ecology & Evolution*, 23(5), pp. 261–267.

Bommarco, R., Kleijn, D. and Potts, S. G. (2013) 'Ecological intensification: Harnessing ecosystem services for food security', *Trends in Ecology & Evolution*, 28(4), pp. 230–238.

Castle, S. E., Miller, D. C., Ordonez, P. J., Baylis, K. and Hughes, K. (2021) 'The impacts of agroforestry interventions on agricultural productivity, ecosystem services, and human well-being in low-and middle-income countries: A systematic review', *Campbell Systematic Reviews*, 17(2), p. e1167.

Chavan, S., Newaj, R., Keerthika, A., Ram, A., Jha, A. and Kumar, A. (2014) 'Agroforestry for adaptation and mitigation of climate change,' *Popular Kheti*, 2, pp. 214–219.

Convention on Biological Diversity. (2016) *Biodiversity and the 2030 agenda for sustainable development*. Technical Note.

Crews, T. E. and Bender, S. F., 2011. *Agroecology in action: Extending alternative agriculture through social networks*. Cambridge: MIT Press.

Datta, A., Bhattacharyya, R., Maity, S. K., Lal, R., Pandey, C. B., Chaturvedi, O. P. and Kundu, S. (2017) 'Impact of long-term cultivation and integrated nutrient management on soil organic carbon stocks and fractions under rainfed agroecosystem', *Journal of Soil Science and Plant Nutrition*, 17(3), pp. 645–657.

Dixon, R. K. (1995) 'Agroforestry systems: Sources of sinks of greenhouse gases?', *Agroforestry Systems*, 31, pp. 99–116.

Dubois, O. (2011). *The state of the world's land and water resources for food and agriculture: Managing systems at risk.* Viale delle Terme di Caracalla, Rome, Italy: The Food and Agriculture Organization of the United Nations and Earthscan. https://openknowledge. fao.org/server/api/core/bitstreams/0dda22d4-41fa-4c16-9090-55ade8cadf66/ content

Dawson, D., VanLandeghem, M. M., Asquith, W. H. and Patiño, R. (2015) 'Long-term trends in reservoir water quality and quantity in two major river basins of the southern Great Plains,' *Lake and Reservoir Management,* 31(3), pp. 254–279.

Ellison, D., Futter, M. N. and Bishop, K. (2017) 'On the forest cover-water yield debate: From demand-to supply-Side thinking', *Global Change Biology,* 23(3), pp. 1007–1023.

FAO. (2020) *Global Forest Resources Assessment 2020 – Key findings. Rome.* https:// doi.org/10.4060/ca8753en

Farley, K. A., Jobbágy, E. G. and Jackson, R. B. (2005) 'Effects of afforestation on water yield: A global synthesis with implications for policy', *Global Change Biology,* 11(10), pp. 1565–1576.

Feliciano, D., Ledo, A., Hillier, J. and Nayak, D. R. (2018) 'Which agroforestry options give the greatest soil and above ground carbon benefits in different world regions?, *Agriculture, Ecosystems & Environment,* 254, pp. 117–129.

Food and Agriculture Organization of the United Nations (FAO). (2020). *The state of the world's forests 2020 - forests, biodiversity and people.* Rome, Italy: FAO. Retrieved from http://www.fao.org/3/ca8642en/online/ca8642en.html

Food Security Indicators. (FAO, 2020). http://www.fao.org/economic/ess/ess-fs/ ess-fadata/en/#.XiYStoh7mcw

Garrity, D. P., Akinnifesi, F. K., Ajayi, O. C., Weldesemayat, S. G., Mowo, J. G., Kalinganire, A. and Bayala, J. (2010) 'Evergreen agriculture: A robust approach to sustainable food security in Africa', *Food Security,* 2, pp. 197–214.

Hailu, T., Negash, L. and Olsson, M. (2000) 'Millettia ferruginea from southern Ethiopia: Impacts on soil fertility and growth of maize', *Agroforestry Systems,* 48(1), p. 9.

Hernandez-Santin, C., Amati, M., Bekessy, S. and Desha, C. (2023) 'Integrating biodiversity as a non-human stakeholder within urban development', *Landscape and Urban Planning,* 232, p. 104678.

Hisano, M., Searle, E. B. and Chen, H. Y. (2018) 'Biodiversity as a solution to mitigate climate change impacts on the functioning of forest ecosystems', *Biological Reviews,* 93(1), pp. 439–456.

Jackson, R. B., Jobbágy, E. G. and Nosetto, M. D., 2005. 'Ecohydrology in a human-dominated landscape: Examples from the United States and Argentina', *Water Resources Research,* 41(3). https://jacksonlab.stanford.edu/sites/g/files/ sbiybj20871/files/media/file/eh09.pdf

Jat, M. L., Dagar, J. C., Sapkota, T. B., Govaerts, B., Ridaura, S. L., Saharawat, Y. S. and Stirling, C. (2016) 'Climate change and agriculture: Adaptation strategies and mitigation opportunities for food security in South Asia and Latin America', *Advances in Agronomy,* 137, pp. 127–235.

Jose, S. (2009) 'Agroforestry for ecosystem services And environmental benefits: An overview', *Agroforestry Systems,* 76, pp. 1–10.

Kar, A., Moharana, P. C., Raina, P., Kumar, M., Soni, M. L., Santra, P. and Dhinwa, P. S. (2009) 'Desertification and its control measures,' n Kar, A., Garg, B. K., Singh, M. P. and Kathju, S. (eds.), *Trends in arid zone research in India.* Jodhpur: Central Arid Zone Research Institute, pp. 1–47. https://www.researchgate.net/ publication/302902688_Desertification_and_its_control_measures

Kay, S., Graves, A., Palma, J. H., Moreno, G., Roces-Díaz, J. V., Aviron, S. and Herzog, F. (2019) 'Agroforestry is paying off–Economic evaluation of ecosystem services in European landscapes with and without agroforestry systems', *Ecosystem Services,* 36, p. 100896.

Kay, S., Rega, C., Moreno, G., den Herder, M., Palma, J. H., Borek, R., ... and Herzog, F. (2019) 'Agroforestry creates carbon sinks whilst enhancing the environment in agricultural landscapes in Europe,' *Land Use Policy, 83*, pp. 581–593.

Kearney, S. P., Fonte, S. J., García, E., Siles, P., Chan, K. M. A. and Smukler, S. M. (2019) 'Evaluating ecosystem service trade-offs and synergies from slash-and-mulch agroforestry systems in el Salvador', *Ecological Indicators*, 105, pp. 264–278.

Kiptot, E. and Franzel, S. (2012) 'Gender and agroforestry in Africa: A review of women's participation', *Agroforestry Systems*, 84, pp. 35–58.

Kumar, S., Bijalwan, A., Singh, B., Rawat, D., Yewale, A. G., Riyal, M. K. and Thakur, T. K. (2021) 'Comparison of carbon sequestration potential of Quercus leucotrichophora–based agroforestry systems and natural forest in Central himalaya, India', *Water, Air, & Soil Pollution*, 232(9), p. 350.

Kumar, M. and Singh, H. (2020). Agroforestry as a nature-based solution for reducing community dependence on forests to safeguard forests in rainfed areas of India. *Nature-Based Solutions for Resilient Ecosystems and Societies.* https://doi.org/10.1007/978-981-15-4712-6_17.

Lasco, R. D., Delfino, R. J. P. and Espaldon, M. L. O. (2014) 'Agroforestry systems: Helping smallholders adapt to climate risks while mitigating climate change', *Wiley Interdisciplinary Reviews: Climate Change*, 5(6), pp. 825–833.

MacKinnon, K., van Ham, C., Reilly, K., & Hopkins, J. (2019). Nature-based solutions and protected areas to improve urban biodiversity and health. Biodiversity and health in the face of climate change, 363-380.https://link.springer.com/content/pdf/10.1007/978-3-030-02318-8_16.pdf

Makundi, W. R. and Sathaye, J. A. (2004) 'GHG mitigation potential and cost in tropical forestry—Relative role for agroforestry,' *Environment, Development and Sustainability*, 6, pp. 235–260. https://doi.org/10.1023/B:ENVI.0000003639.47214.8c.

Marselle, M. R., Hartig, T., Cox, D. T., De Bell, S., Knapp, S., Lindley, S. and Bonn, A. (2021) 'Pathways linking biodiversity to human health: A conceptual framework', *Environment International*, 150, p. 106420.

Mbow, C., Van Noordwijk, M., Luedeling, E., Neufeldt, H., Minang, P. A. and Kowero, G. (2014) 'Agroforestry solutions to address food security and climate change challenges in Africa', *Current Opinion in Environmental Sustainability*, 6, pp. 61–67.

Milder, J. C., Hart, A. K., Dobie, P., Minai, J., Zaleski, C. and Uphoff, N. (2010) *Integrated landscape management for agriculture, rural livelihoods, and ecosystem conservation: An assessment of experience from Latin America and the Caribbean.* The World Bank. https://www.researchgate.net/publication/262767896_Integrated_landscape_management_for_agriculture_rural_livelihoods_and_ecosystem_conservation_An_assessment_of_experience_from_Latin_America_and_the_Caribbean

Miñarro, M. and Prida, E. (2013) 'Hedgerows surrounding organic apple orchards in north-west s pain: Potential to conserve beneficial insects', *Agricultural and Forest Entomology*, 15(4), pp. 382–390.

Muhie, S. H. (2022). 'Novel approaches and practices to sustainable agriculture', *Journal of Agriculture and Food Research*, 10, p. 100446.

Nair, P. K. (2012) 'The coming of age of agroforestry', *Journal of the Science of Food and Agriculture*, 92(10), pp. 1849–1851.

Newaj, R., Chavan, S. and Prasad, R. (2013) 'Agroforestry as a strategy for climate change adaptation and mitigation', *Indian Journal of Agroforestry*, 15(2), pp. 41–48.

Nicholls, C. I. and Altieri, M. A. (2013) 'Plant biodiversity enhances bees and other insect pollinators in agroecosystems. A review', *Agronomy for Sustainable Development*, 33, pp. 257–274.

Olsson, L., H. Barbosa, S. Bhadwal, A. Cowie, K. Delusca, D. Flores-Renteria, K. Hermans, E. Jobbagy, W. Kurz, D. Li, D.J. Sonwa, L. Stringer, 2019: Land Degradation. In: Climate Change and Land: an IPCC special report on climate change, desertification, land degradation, sustainable land management, food security, and greenhouse gas fluxes in terrestrial ecosystems [P.R. Shukla, J. Skea, E. Calvo Buendia, V. Masson-Delmotte, H.-O. Pörtner, D. C. Roberts, P. Zhai, R. Slade, S. Connors, R. van Diemen, M. Ferrat, E. Haughey, S. Luz, S. Neogi, M. Pathak, J. Petzold, J. Portugal Pereira, P. Vyas, E. Huntley, K. Kissick, M. Belkacemi, J. Malley, (eds.)]. In press.https://www.ipcc.ch/site/assets/uploads/sites/4/2019/11/07_Chapter-4.pdf

Perfecto, I., Vandermeer, J. and Wright, A. L. (2009) *Nature's matrix: Linking agriculture, conservation, and food sovereignty*. Routledge. https://www.routledge.com/Natures-Matrix-Linking-Agriculture-Biodiversity-Conservation-and-Food-Sovereignty/Perfecto-Vandermeer-Wright/p/book/9780367137816?srsltid=AfmBOoo3H6x8tGoFm8Njfm-KuimSIJmrwgxMFmt8IQ7PS0nyuTRpoZ-L

Rakotovao, N. H., Rasoarinaivo, A. R., Razafimbelo, T., Blanchart, E. and Albrecht, A. (2022) 'Organic inputs in agroforestry systems improve soil organic carbon storage in Itasy, Madagascar', *Regional Environmental Change*, 22, pp. 1–13.

Saha, A., Ramya, V. L., Jesna, P. K. *et al.* (2021) 'Evaluation of spatio-temporal changes in surface water quality and their suitability for designated uses, Mettur Reservoir, India,' *Natural Resource Reserve*, 30, pp. 1367–1394. https://doi.org/10.1007/s11053-020-09790-5

Sahoo, G. and Wani, A. M. (2019) 'Multifunctional agroforestry systems in India for livelihoods', *Annals of Horticulture*, 12(2), pp. 139–149.

Sathaye, J. A. and Ravindranath, N. H. (1998) 'Climate change mitigation in the energy and forestry sectors of developing countries', *Annual Review of Energy and the Environment*, 23(1), pp. 387–437.

Sileshi, G. W., Mafongoya, P. L. and Nath, A. J. (2020) 'Agroforestry systems for improving nutrient recycling and soil fertility on degraded lands', *Agroforestry for Degraded Landscapes: Recent Advances and Emerging Challenges*, 1, pp. 225–253.

Tchoundjeu, Z., Degrande, A., Leakey, R. R., Nimino, G., Kemajou, E., Asaah, E. and Tsobeng, A. (2010) 'Impacts of participatory tree domestication on farmer livelihoods in West and Central Africa', *Forests, Trees and Livelihoods*, 19(3), pp. 217–234.

Telwala, Y. (2023) 'Unlocking the potential of agroforestry as a nature-based solution for localizing sustainable development goals: A case study from a drought-prone region in rural India', *Nature-Based Solutions*, 3, p. 100045.

Torralba, M., Fagerholm, N., Burgess, P. J., Moreno, G. and Plieninger, T. (2016) 'Do European agroforestry systems enhance biodiversity and ecosystem services? A meta-analysis', *Agriculture, Ecosystems & Environment*, 230, pp. 150–161.

Tscharntke, T., Tylianakis, J. M., Rand, T. A., Didham, R. K., Fahrig, L., Batáry, P., Bengtsson, J., Clough, Y., Crist, T. O., Dormann, C. F. and Ewers, R. M. (2012) 'Landscape moderation of biodiversity patterns and processes—Eight hypotheses', *Biological Reviews*, 87(3), pp. 661–685.

United Nations (2017) 'World population prospects: The 2017 revision, key findings and advance tables,' Department of Economics and Social Affairs PD, editor. New York: United Nations, 46(1).

United Nations (2021) 'The sustainable development goals report 2021,' United Nations Publication Issued by the Department of Economic and Social Affairs, 64. https://www.un.org/en/desa/sustainable-development-goals-sdgs

van Noordwijk, M., Lusiana, B., Hairiah, K., Velarde, S. J. and Pradhan, U. (2014) 'Opportunities for joint coffee and environmental management: Insights from a marginal landscape in Indonesia', *Agroforestry Systems*, 88(5), pp. 809–823.

Wartman, P., Van Acker, R. and Martin, R. C. (2018) 'Temperate agroforestry: How forest garden systems combined with people-based ethics can transform culture', *Sustainability*, 10(7), p. 2246.

World Bank. (2020). *Trading for development in the age of global value chains.* World Bank. https://openknowledge.worldbank.org/bitstream/handle/10986/32437/9781464814570.pdf

Zhu, X., Liu, W., Jiang, X. J., Wang, P. and Li, W. (2018) 'Effects of land-use changes on runoff and sediment yield: Implications for soil conservation and forest management in Xishuangbanna, southwest China', *Land Degradation & Development*, 29(9), pp. 2962–2974.

Zomer, R. J., Neufeldt, H., Xu, J., Ahrends, A., Bossio, D., Trabucco, A. and Wang, M. (2016) 'Global tree cover and biomass carbon on agricultural land: The contribution of agroforestry to global and national carbon budgets', *Scientific Reports*, 6(1), p. 29987.

Index